Senses and Sensibilities

JILLYN SMITH

Illustrated by Chris Smith

Wiley Science Editions

JOHN WILEY & SONS, INC.

New York • Chichester • Brisbane • Toronto • Singapore

For the memory of Mom,
who loved to laugh,
and for Dad, Jean, and Joanie,
who love stories.

Publisher: Stephen Kippur
Editor: David Sobel
Managing Editor: Corinne McCormick
Editing, Design, and Production: G&H SOHO, Ltd.

Library of Congress Cataloging-in-Publication Data

Smith, Jillyn.
Senses and sensibilities.

(Wiley science editions)
Bibliography: p.
1. Senses and sensation. I. Title. II. Series.
QP431.S54 1989 612′.8 88-33972
ISBN 0-471-50657-5
ISBN 0-471-61839-X (pbk.)

Printed in the United States of America

89 90 10 9 8 7 6 5 4 3 2 1

Prologue

Last night I took a bath. I found a recording of Beethoven's Ninth Symphony to play on the stereo, and I turned it up, loud. I sprinkled the bathwater with a few drops of wood spice oil, and I took with me four date bars and a book of short stories. When the water cooled, I turned on more hot water with my foot. I read four short stories in the bathtub, one for each date bar.

This morning, early, I sit at my kitchen table, still faintly smelling of wood spice, writing sentences about the senses. The coffeepot stops gurgling, a signal I've been waiting for. I pour a cup and add cream, judging by the color change when I've added the right amount. I've come to crave the smell and taste of coffee early in the morning. Lately, however, I've omitted sugar—it was beginning to taste too sweet. I'm also finding cola drinks sweet these days, an objection I had heard in the past from others. At work I often choose orange juice from snack machines. I wonder if my taste buds are changing, if my body is trying to tell me not to consume so much sugar or whether I might have a need for vitamin C.

Mr. and Mrs. Beeper, the zebra finches who live in a cage near my living room window, are awake early, too, singing their duet in the morning light. In the afternoon, many-faceted crystals hanging from the west window will break the light into hundreds of little dancing rainbows. Sometimes the Beepers see themselves in the mirror of their hanging toy, and they peck on it and the bell tinkles. Later they will sing with the music that comes from the radio.

Cruiser, the one-eyed garbage-can cat who adopted me and follows me everywhere, has eaten his breakfast of dry cat food and naps upon the bench beside me. I stroke his coat, which is alive with warmth and electricity. Once deprived, Cruiser is now overweight and has a problem common to the overweight. He snores, adding an irregular sound to the blowing of the furnace fan and the ticking of the clock.

It is almost March, and the morning sun is shining earlier, higher, and warmer through the windows. Patches of green are beginning to appear on the lawn, and the houseplants seem greener and perkier too. Because I am wearing contact lenses, I see clearly across the expanse of the valley. The steep Wellsville Mountains have sharp edges and appear flat, as if they have been painted on the backdrop of blue sky.

I remember my high school English teacher, Mr. Osborn, and how he loved reading and reciting the sentences of Shakespeare and Kipling. I remember how, in class, I diagrammed sentences and identified nouns, verbs, prepositions, and yet how clumsily I created sentences of my own. In my mind I see Mr. Osborn, exasperated with us, teenagers getting used to our circulating hormones, daydreaming about lunch or the opposite sex, writing pitiful, pitiable English "themes."

I hear him. "Observe, observe," Mr. Osborn pleaded with us, as we looked away or doodled. "How can you describe something if you don't look at it, hear it, touch it, taste it, smell it?"

How, indeed?

Acknowledgments

In the early 1970s I studied zoology at Arizona State University. My favorite courses were those taught by animal behaviorist John Alcock. His questions and enthusiasm were nourishment that he and his wife, Sue, often augmented with food-filled socials at their home (also important for starving graduate students). John challenged his students to look at the living world in terms of optimality theory and adaptive significance, to think creatively about what we observed, to ask the why questions as well as what and how questions. By personal example John also encourages lively writing about the science of life.

In the early 1980s, after a decade of trying to vivify biology for high school and college students, I studied journalism at the University of Arizona. My UA professors, newsmen all—Phil Mangelsdorf, Jim Patten, Pete Potter, Steve Emerine, Bill Greer, Don Carson—demanded that I write clear and accurate prose. They often invoked the mother principles of clarity and accuracy: "Write for your mother," they said, and, "Check everything. If your mother says she loves you, check it out." They were ruthless in their critiques and generous in their belief that I could improve. Phil finally admitted that he liked my paper on hearing disorders, which has provided the seed crystal for this book.

I love them all, and I thank them.

I also thank my friends and colleagues, the scientists at Utah State University, where I work, and the professionals at Logan Regional Hospital and elsewhere who listened to me, told me stories, brought me books and articles, and shared their insights and knowledge. Just some of my riches: Edie and Mike Allen, J. R. Allred, Diane Browning, Ellen Caldwell, Kay Camperell, John Carlisle, Carl Cheney, Helene Cohen, Bob Cole, Veronica Dougherty, Kit Flannery, Jim Fuller, Debbie Gessaman, Marilynne Glatfelter, Susan

Grey, Kurt Gutknecht,Dennis Hinkamp, Chuck Lent, Jim Lowell, Ivan Palmblad, Karrie Pennington, Jeanne Pietig, Scott Raymond, Nancy Rosenblatt, Kris Saunders, Scot Smith, Spencer Smith, Sheryl Spriet, Gene Sturla, Jan Tucker, Rory Weaver, Alexa West, Leona Windley (who said, "You may not have any friends by the time you finish this book"), Steve Winitsky, and Maggy Zanger.

Many scientists, known to me only by their clever work, generously sent reprints and other information.

John Alcock, Ron Daines, Derek Davenport, Tom Lyon, Julie Simon, and John Wood read embryonic portions of the manuscript, asked questions, and made suggestions. Ken Brewer's comments were especially helpful. My editor, David Sobel, judicious in his use of such words as *stern* and *terrific*, mothered me throughout the project, invoking the mother principles from time to time. He and his staff are terrific. David Sassian, Corinne McCormick, and Nancy Woodruff were extremely genial to work with.

Chris Smith came from Tucson to Logan to work on illustrations, bringing her magic pens and effervescence along with seaweed and adzuki beans. Her life is art, and I treasure our friendship. Ron Shook, another dietary variant, helped with discussion, proofreading, and entertainment while consuming diet soda, sunflower seeds, chocolate bars, and an occasional Twinkie. Thanks, Ron.

Three biologists and former teachers are gone now but left me with some good contradictory life instructions: Gordon Castle said, "Be organized and prepared." Leon Jordan said, "Don't tell the students everything you know." And E. Raymond Hall said, "There comes a time you have to go to press." I owe a great deal of who I am to them and to my former biology students, who have provided abundant memories.

Finally, I typed the manuscript myself, with limited help from the cat, the computer gods, and Mozart's *Requiem*. Diane said I should have tried *The Magic Flute*. She also said, "Lace is beautiful because of the holes."

Contents

The Senses: Peepholes on the World 1
Environmental Receivers 2
A Long History of Sensing 6
Head and Hands 12
Hearing: The Noblest Faculty 15
The Ear and Sound 16
Jaws into Ears, or Installation of the Equipment 22
The Human Voice and Music 24
Different Wavelengths 27
Children, Language, and Ears 33
Audiograms 35
Hearing Loss 41
The Noblest Faculty 44
Problems of Detection and Paranoia 46
Fluent on Their Fingers 47
Aids, Tickle Belts, and Bionic Ears 49
The Sense of Communication 51
Vision: A Viewpoint 53
The Eyes: Extensions of the Brain 54
The Exciting Visible Spectrum of Light 59
I See, I See 65
Color Defects 70
The Eye as Camera, and Optograms 73
We See What We See 75
Lenses by Nature and Humans 77
Vision Loss 82

Smaller and Larger 85
Tears 87
Dilation with Pleasure 88
Old Eyebrow Greetings and New Blink Science 89
Vision, Television 91

Smell: The Inarticulate Sense 93
Fishing for Chemicals 94
An Ancient Sense 98
Classifying Smells 106
Pheromones: Within-Species Coercion 112
Sex Smells 117
Scratching and Sniffing 118
Smell Memories 121
Odor Blindness, Anosmia, Hyposmia 123
Artificial Noses 126
Environmental Odors 126
People Smells 127
Why Do Humans Wash? 129
Perfumery 130
A Life of Odors 132

Taste: Accounting for It 135
The Tongue and Taste 136
From Greek Tongues to Psychophysics 142
Sweet, Sweet Sugar 146
Sugar as Mood Food 151
Fooling the Tongue: Sugar Substitutes 152
Sour, Salty 157
Bitter, Hot, Cold, and New 162
Differences of Tongues: Season to Taste 165
The Fine Art of Tasting 168
Taste Disorders: Dysgeusia 171
Otherworldly Tasting, or the Fly's Feet 173
Taste Aversions 175
Keeping up with Taste 176

Touch: The Confirmatory Sense 179
Sensory Enfoldment 180
Hands: The Organs of Touch? 186
Temperature Monitors 189
Pain: Hurtful, Warning 191
Social Contact 196
Touch and Development 198
Therapeutic Touch 200
Gratuitous Touch 201
To Be Is to Touch 202

Seeking Sensory Experience 205
Human Versatility 206
General Sensitivity 207
Adaptation and Recovery 209
Feed Me, Feed Me: Superstimulation 211
Of Spice and Space 212

References 215
Index 225

For there is no conception in a man's mind, which hath not first been begotten upon the organs of Sense.

—THOMAS HOBBES

(Touch, while singing):
Head, shoulders, knees, and toes,
Head, shoulders, knees, and toes,
Eyes, ears, mouth, and nose,
That's the way the story goes.

—MY KINDERGARTEN TEACHER

O · N · E

▼

The Senses: Peepholes on the World

Environmental Receivers

"I have caught life," says a Kurt Vonnegut character. "I have come down with life. I was a wisp of undifferentiated nothingness, and then a little peephole opened quite suddenly. Voices began to describe me and my surroundings."

Life. Peepholes. The peepholes of life. Senses to detect the world. Not just vision, that is, light sensation, but peepholes of mechanical sensation—hearing and touch—and peepholes of chemosensation—taste and smell. The peepholes begin to open before birth. They are synonymous with life.

Hearing, which Ludwig van Beethoven called "the noblest faculty," may be the first peephole to open, after the general sensitivity of the skin. A fetus begins to sense the beat of its mother's heart and to detect the mechanical rhythm of her breathing. This internal rhythm we all hear and have within ourselves is duplicated in other aspects of our lives, such as the iambic (ta-TUM) cadence of our speech, walking, dancing, music, poetry. It has been suggested that children begin to speak in double syllables—da-da, ma-ma—in imitation of the paired heartbeat sounds, and that a child instinctively clings to the left side, the heartbeat side, of the mother's breast.

The womb is a dark place. Eyes are still forming, and the neuronal connections of the brain that interpret and integrate information from the senses continue to develop after birth in interaction with their environment. The eyes of all newborn humans are dark and nondescript in color, farsighted, and wobbly. They later develop the iris pigments that filter light and eliminate fuzziness, the muscle control for focusing, and stereoscopic or depth vision.

The sense of taste is partially developed in the womb. The sugar-detecting ability develops before birth, and a baby will reflexively suck a sugar-containing solution. New babies will react to odors with arousal, and the sense of smell may begin to be of significance as the suckling baby recognizes the comforting odor of secretions from areolar glands on the mother's nipples.

The baby's smooth skin is stimulated during birth by uterine contractions and after birth by the pressure and warmth of proximate hands and bodies. The sensory structures and capacities, and the brain's ability to interpret sensory signals, develop in tandem. Stimulation of the senses in turn stimulates nervous system cells to grow and to make meaningful contacts. By stimulation of the peepholes of the senses, we become more sensitive. We become human, complete. Hearing helps develop the template of language. Early stimulation of the visual system reinforces itself. Touch confirms the visual sense. Chemosensations provide more experience of the world. Peepholes open wider and wider. Peepholes connect us with each other, and we learn, storing information and experiences for later reference.

Throughout our lives, until the peepholes of the senses close, we are sentient beings, sensing the energy and chemicals and objects in our environments. But we are complicated. Sensation is different from perception. Sensation can be described more easily; perception refers to the sense our brains make of sensation.

The information picked up by our antennae to the world, our senses, is converted into a form that the brain can integrate, compare, store, and interpret. We have learned a great deal about the structure and function of our sensory apparatus: the "special" senses of hearing, vision, taste, and smell and the more "general" senses distributed over the skin (special, too). But the brain's processing, decoding, integration, and retention of information gleaned from the environment via the senses still holds much mystery. The human brain I am using as I write, that three pounds of gray matter in my skull, represents biology's grand frontier.

Because our species shares genes, genetic instructions, that say "human," we have our general equipment for sensing the world in common with other humans. Variation is also a universal. It is our differing individual sensations, perceptions, and associated experiences that form us and make us who we are. We are the products of our genetic makeup, our resultant physiology, our environment and our experience. Age makes a difference too.

Our senses allow us a wide variety of experience: to appreciate human voices and a symphony, for instance, as well as the screech of brakes or the whistle of a train—coming and going. They allow us to distinguish ripe berries and the vivid colors of a van Gogh painting as well as the stars in the night sky and seats within a dark movie theater. They allow us to detect the aromatics created by chemical reactions in plants and animals, including ourselves, and the short chains of volatile amino acids released from rotting flesh. They allow us to discriminate among bitter, sweet, sour, and salty tastes and to combine the information. They allow us to derive pleasure from touching a cat's fur, a piece of silk, or a baby's smooth skin and to protect ourselves from potential danger—sources of heat, cold, pain, or crushing pressure. Through ingeniously engineered structures and specialized nervous system cells, called *receptors*, and through the electrochemical transmission in our nervous systems—messages traveling electrically along the neurons, leaping chemically from neuron to neuron—our sensations provide the experience of life.

Our ears trap the mechanical energy of sound waves and condense them while cleverly keeping and sorting the information the waves contain. The most important sounds humans hear are those made by other humans, and we are especially keen at detecting the nuances of human speech. We hear others, but we also hear ourselves, which allows us to check and modify our own transmissions. The ear contains organs of equilibrium and balance, too, to tell us we are right side up; their function is more ancient than is hearing.

Our eyes are sensitive to light energy and help convert it into mental pictures of the world. Taste and smell, results of specialized receptors in the mouth and nasal passages, tell us about the chemical world in which we find ourselves—close by and at a distance. The sense of touch, by which we detect pressure, temperature, and pain, is spread over the body but is especially rich in the hands and face, those parts of the body that encounter the environment first and foremost. Touch is also rich in the parts of the body significant in sexual context and contact, playing a significant role in bonding and reproduction.

For most of human history, the special senses were our only detection mechanisms for the reality of the external environment. Only comparatively recently in human history have we developed instruments to sense more, to see beyond the limits of the human eye both macroscopically and microscopically, to hear beyond the limits of the human ear, to amplify stimuli, and to analyze chemicals molecule by molecule. These extensions of the human senses have given us glimpses of a broader and deeper universe, expanding our reality.

Although we each have an overall unique sensory experience, we also have many experiences that we share with others. If I mention fresh-baked bread, others with similar experience will remember similar pleasant sensations. In our minds, we see the bread, we smell it, we taste it, we touch it. We have some universal pleasant experiences and some universal "phews."

Because we are social, we are constantly gathering information about others of our species. We look into each other's eyes, we whisper into each other's ears, we cup our ears, we watch mouths for smiles to indicate pleasure and noses for wrinkles to indicate displeasure. We smell each other and make smell associations and memories. We touch each other and we feel the touch of others upon us and their warmth. We seek contact. We need hugs.

As we often discover and rediscover to our surprise or delight, we are all tuned to receive slightly different information because of our individual differences in genetics, experience, age, and sex. My friend Gene's strategy for birdwatching is different from mine because he is color-blind. He pays attention to other clues: silhouettes and calls. My friend John often has to be told his wristwatch alarm is beeping, because he can't hear it. My friend Leona uses artificial sweeteners in all her drinks; I dislike them because of the bitter aftertaste. My friend Kris mentions that sodium pentathol, a common surgical anesthetic, tastes like onions. My friend Ron says musk and sweat are identical smells. I find them as different as new shoes and Tabu perfume.

Our planet-mates, the other animals, have even different tuning, related to their particular ecology and evolutionary history and to the way their ancestors have exploited the environment. Just as our senses make sense for us, the senses of the other animals make biological sense for the lives they lead, the niches they fill. Bats hear moths, moths hear bats, barn owls hear scampering mice, frogs' eyes are "bug detectors," ticks detect butyric acid emanating from mammals close by, turkey vultures smell fresh carcasses, snakes and lizards smell the air with their tongues.

Some of the most clever and ingenious sense organs are found within the insects, whose sensory abilities are as highly developed as those of the vertebrates. Grasshoppers hear with their legs, moths and mosquitoes smell with their antennae, blowflies and butterflies taste with their feet. Dragonflies, with wraparound eyes, see a picture of almost 360 degrees (and are understandably difficult to catch).

We are far from alone on Earth. Understanding the peepholes of Earth's other animals can broaden our own perspectives. Tapping into their worlds can have its practical and aesthetic applications, placing our own sensory apparatus into better context.

Although we have deciphered much about our sensations and perceptions, especially in the last century or two, and have developed cunning ways of study and measurement, experimental findings are often piecemeal, puzzling, intriguing, and unexpected. Just as we have come to understand some of the normal human brain function by observing the effects of brain damage and what happens when a portion of the brain is no longer functioning, so also have we come to understand something about the human senses from observations of defects, diseases, or injuries that affect the special senses and alter sensory systems and perceptions. Humans have also made strides in helping those with sensory deficits and in developing aids for failing senses.

A Long History of Sensing

When I was a high school biology student, the chapter on unifying principles was the final one in the textbook, or maybe it was an appendix, along with the metric system, which wasn't terribly useful at the time either. Of course we never reached the end of the book, but we kept busy enough, learning the parts of an eye, the parts of a leaf, the parts of a flower (girls wear the sexual structures of plants to the prom, I realized gleefully). In retrospect, I think that final chapter, that appendix, must have mentioned *evolution*, an important word that means change, something that life does best, has always done best. Here I mention it first, before I get to the details.

The way humans are today is a result of our evolutionary history, and the same is true of all the other designs we see for living things. Our species is just one of the millions of kinds of life on the planet, and a recent design at that. Life has a long intimacy with Earth. It has survived and reproduced and evolved here for more than 3 billion years. For the majority of that time, life was single-celled and aquatic, yet not exactly simple. Multicelled life forms, and especially terrestrial forms, are relatively new developments in the long chain of life. Some life forms, such as bacteria and algae that inhabit the Earth with us today, are ancient successful designs and apparently have changed little. Sharks, although much younger than algae, are still an old, old blueprint. For other forms—particularly the later large land-dwelling groups of animals, the insects and vertebrates—change and proliferation have been the rule. Their elaboration over evolutionary time has been rapid and terribly clever, in response to the demands of the capricious land environment, which is more changeable than water.

Yet, within the diversity of life, among the humans and the bats and the fruit flies, is unity—the unity of the genetic material DNA, the unity of cells, and the unity of the basic necessities for life. These basic necessities are variously called surviving and reproducing, hunger and sex, feeding and avoidance of predation, or finding prey or accumulating resources while avoiding being prey. To meet the needs of life—survival and reproduction—all living things need information. We have in common with our planet-mates the necessity for gathering information about the environment and of communicating with our own species.

Perhaps the oldest kind of communication is chemical. Chemical signals can come from both the nonliving and living environment, but detection of other living things is most critical and the chemical senses reflect this in being best at detecting organic molecules. It's easy for a watery bag of chemicals, a living thing, to leave a few chemicals around and to react to other chemicals chemically. Primitive bacteria and protozoans can readily distinguish among different chemicals, as they show by their feeding and avoidance behavior. Chemicals must have played a critical role in the recognition of cells of the same species in primitive organisms. And chemicals must have played a critical role in the development of sexual reproduction in life's watery womb, flagging egg and sperm cells so they could properly signal and detect each other.

In addition to the chemical environment, living things have developed and evolved in the physical environment of an Earth that rotates as it revolves around the sun, ergo, cycles of light and dark and temperature changes. Life has evolved in an environment with electricity and gravity and physical movement of earth, air, and water. Living things have developed sensory mechanisms that also detect these kinds of stimuli.

As animals evolved into multicelled forms, division of labor and specialization of cells were possible. Primitive multicelled animals, such as sponges and jellyfish, are asymmetrical or pie shaped. The evolution and elaboration of nervous systems, of which the special senses are a part, was accompanied in higher animals by the development of a head end where sensory structures were concentrated and of bilateral, or two-sided, symmetry. In bilaterally symmetrical animals such as worms, fruit flies, lizards, and ourselves, the sides are mirror images of each other. So are the sensory structures.

The bilaterally symmetrical multicelled animal, with a head and tail end, has been a good and lasting design for life on Earth. The head end proceeds foremost, gathering information about the environment. In the more advanced animals, paired auditory organs allow location of sounds by triangulation because sound energy arrives at the two organs at slightly different times. Paired eyes, with the potential for overlapping fields of vision, promote depth perception.

Each species has sensory organs suited to its particular needs and its particular exploitation of the environment. Animals, including ourselves, are products of selective forces acting on their ancestors, favoring those best suited to survive and reproduce in their particular environments.

For answers to questions about the adaptive significance of traits, we must look to the past. Living things lag a little in adaptation to their environments. They are a product of ancestors who were successful at survival and reproduction, ancestors who passed on those traits. In each living thing, including ourselves, we find pieces of past lives.

Natural selection works on individual animals. With each generation, the genes are shuffled, creating new combinations. Mutation also provides raw material for selection to work on. A trait that confers a survival and reproductive advantage to an individual, however slight, eventually replaces less advantageous traits over evolutionary time. The frog with eyes that respond to a small moving object will catch more flies, faster, and can spend more time courting and mating with females than a frog with less sensitive vision. A rat that associates a certain odor or taste with a plant that evoked a past illness will have an advantage over a rat that makes no distinction and wastes time eating poison. The first rat will be healthier and have more time to spend on reproduction. A fruit fly male that recognizes a receptive mate of the same species will not waste time courting the wrong female. A fruit fly female that responds to a chemical indicating a good food source for her developing larvae will have an advantage over the fruit fly that lacks the trait.

The result of evolutionary history is that living things don't waste energy on characteristics that have no meaning for them (or that had no meaning for their ancestors, whence they came). Biologists call this idea *optimality theory*, and looking through the lens of optimality theory helps us make sense of the living world. When we observe that all barn owls have one ear lower than the other, we can assume that this peculiar arrangement somehow has contributed to the barn owl's success as a life form. If a rat's nose seems more elaborate and sensitive than it needs to be for food getting, we can ask whether it might have another function, perhaps that of recognizing the individual smells of other rats.

Because we are humans, and because we have incredibly complicated our lives with culture, some of the most interesting *why* questions are about ourselves. Why do we like hot baths? Why are we good at recognizing a voice on the telephone after just a hello? Why do we like artificial sweeteners? Perfumes? Television? Rock concerts? How can these questions, many of them about modern life, be related to the demands of our evolutionary past?

Humans are vertebrates. We share certain characteristics with other back-boned animals. We are also mammals, with characteristic mammalian teeth (incisors, canines, premolars, molars), hair,

Most of human evolutionary history has been spent in a hunting-gathering life-style on the savanna. We find the savanna landscape plan appealing and tend to re-create it as we alter the landscapes around us. Humans spend an inordinate amount of time maintaining artificial grasslands. Bonsai trees, some speculate, represent creation of miniature savanna trees.

and mammary glands. We are primates, with the primate adaptations of flattened nails, opposable thumbs, binocular vision (good for judging depth and distance in trees), and color vision (good for detecting fruit). We are of the family Hominidae, genus *Homo*, species *sapiens*, subspecies *sapiens*. (We named ourselves, of course, "doubly thinking.") We are bipedal, with hands that are exquisite manipulators. We are gregarious, social, "groupies." We have well-developed facial muscles, making us "expressive." We have well-developed language, and we like to use it. We find it easy to drive a car, an activity that has requirements similar to navigating through trees and anticipating oncoming branches.

About 4 million years ago, the ancestors of humans moved from a forest environment to an environment of grasses and trees, the savanna. This move was a big one, representing exploitation of a niche not exploited by other primates. On the savanna, humanoids were omnivores. Most of human evolutionary history has been spent in a life-style of hunting and gathering. We became the genus *Homo* about a million years ago. We invented agriculture and a settled existence 8,000 to 11,000 years ago. The Industrial Revolution is little more than two centuries old. We developed flying machines less than a century ago. The space age is 30 years old. We have accelerated change and continue to do so.

But we can't escape our collective past. Our mammalian/primate heritage was molded for several million years by the hunting-and-gathering life of the savanna. Selection for success in that life-style shaped our sensory worlds and preadapted, or prepared, us for the more recent life-styles we have created. Our adaptations for hunting and gathering just happen to be useful currently in industrial and technological society.

Our brains are large and complicated, as a combination of selective factors, including the activities of tool making and use, hunting, cooperation, sociability, aggression, language, planning, and learning. One factor in our development was probably storytelling—relating events in the history of the group. We do love stories. In all these activities, sensory acuity is important.

Human males and females differ physiologically. Adult males in general are larger, with greater upper body strength and greater muscle mass. Females are physiologically suited to bear and suckle young; they have a greater percentage of body fat. Hormonally, adult human males and females live in different universes. What about our sensory universes?

Natural philosophers have struggled with questions about sensations and about perceptions of reality since at least the time of Aristotle (or earlier, with the Chinese), when the seeds of scientific thinking began to sprout. The early natural philosophers shared an ignorance of anatomy, which makes their physiological explanations fantastic, but where physiology and psychology come together, the data have changed less and they still have contributions to make. Humans like to think. We are conscious of the universe we perceive. One line of thought goes so far as to assert that creatures that perceive, by their existence, make the universe exist. As Galileo said, "Tastes, odors, colors and so on are no more than mere names . . . and . . . they reside only in the consciousness."

It's more difficult to do physics than to drive a car, but humans also do physics, and they build instruments to probe deeper into the worlds of the small and the large. As we learn more about the universe, our everyday sensory perceptions fail us in providing appropriate analogies to explain them. The world is no longer a commonsense realm, but curiouser and curiouser. Our brains were developed for coping with the peepholes of common experience, for defining the natural world in our size and time frames, but they also have the ability to chip away at sensing the wider universe, to try to explain the worlds of quarks and black holes in a universe that seems to have no bedrock. Some of our species, today's equivalent of the natural philosophers, spend their lives catching ideas in the nets of their brains and telling others about them.

Head and Hands

By separating our study of the senses, we can better understand them. But nothing in the environment is the private property of any one sense. The senses operate together, and the environmental information that gives us the picture of our world ultimately belongs to the brain, to do with what it will. We can do more than one thing at once, but our brains are selective about the information they take in, about the information we pay attention to. The brain filters, integrates, decides what is important, and lets us know. We can become absorbed in a task and be oblivious to everything except that task.

Understandably, a great deal of our brain is devoted to the reception and analysis of sensory information. Psychologists are fond of the homunculus ("little man") as a graphic representation of the amount of our brains devoted to sensory input compared with the amount devoted to movement of body parts. The homunculus is big headed and big handed. Significantly, the genitalia are also well endowed with sensory receivers.

We do know that we don't sense and perceive everything in our environment, but what we perceive has, and has had, importance for us. In evolutionary terms, our perception of the world, like that of the other life forms, has been tested by the ultimate reality: the ability to find food and to reproduce, resulting in the passing on of the genetic basis for successful designs for sensing the world.

For simplicity's sake, I have focused on what are known classically as the five senses: hearing, vision, taste, smell, and touch.

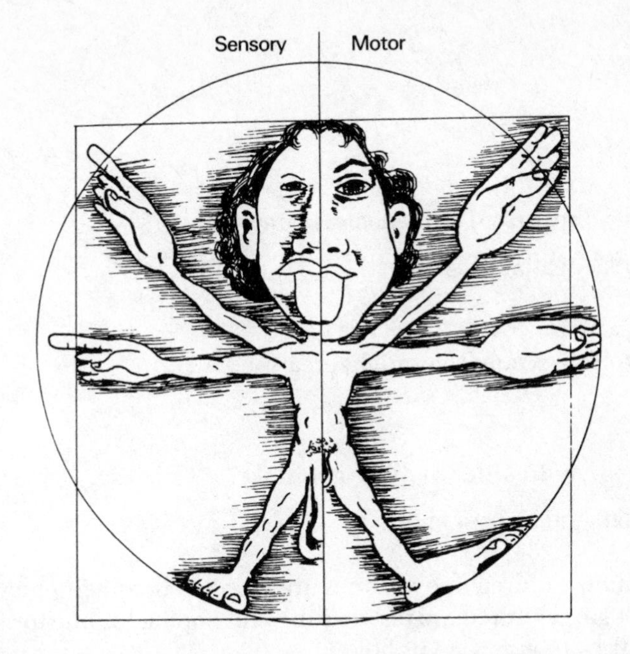

The homunculus or "little man" is used in various ways to graphically represent the relative dedication of the human brain to the senses (input) and to movement of body parts (output). A large portion of the brain is devoted to the hands, face, and genitalia. The lips and genitalia are even more sensory than mobile, and the eyes and hands are given more motor than sensory space in the brain despite our great dependence on them for sensory information. (From Biology: The Science of Life, *by R. A. Wallace, J. L. King, and G. P. Saunders, © 1981. Adapted with permission from Random House, Inc.)*

These are the senses humans are most concerned with (if I were a fish, I would add electroreception). With each sense I have included some information about the origin, structure, and function of the system, about attempts to understand the particular sensation, and about what we can learn from a deficient or dysfunctional sense. A mutation or an individual extreme is always a source of fascination, as is individual experience. As a teacher I learned that my students loved questions of anomaly ("My brother is color-blind. Why can't he see red socks?")

For a peep into different sensory peepholes, I have included some examples of what we know of some of the senses in other animals as well as what they might mean for them. There is much to know and respect about our fellow travelers on the planet Earth. We owe it to ourselves to become better acquainted.

Bid me discourse. I will enchant thine ear.

—William Shakespeare

I was all ear
And took in strains that might create a soul.

—John Milton

Without music, life would be a mistake.

—Friedrich Nietzsche

Listening children know stories are *there*. When their elders sit and begin, children are just waiting and hoping for one to come out, like a mouse from its hole.

—Eudora Welty

Little pitchers have big ears.

—My mother

T · W · O

▼

Hearing: The Noblest Faculty

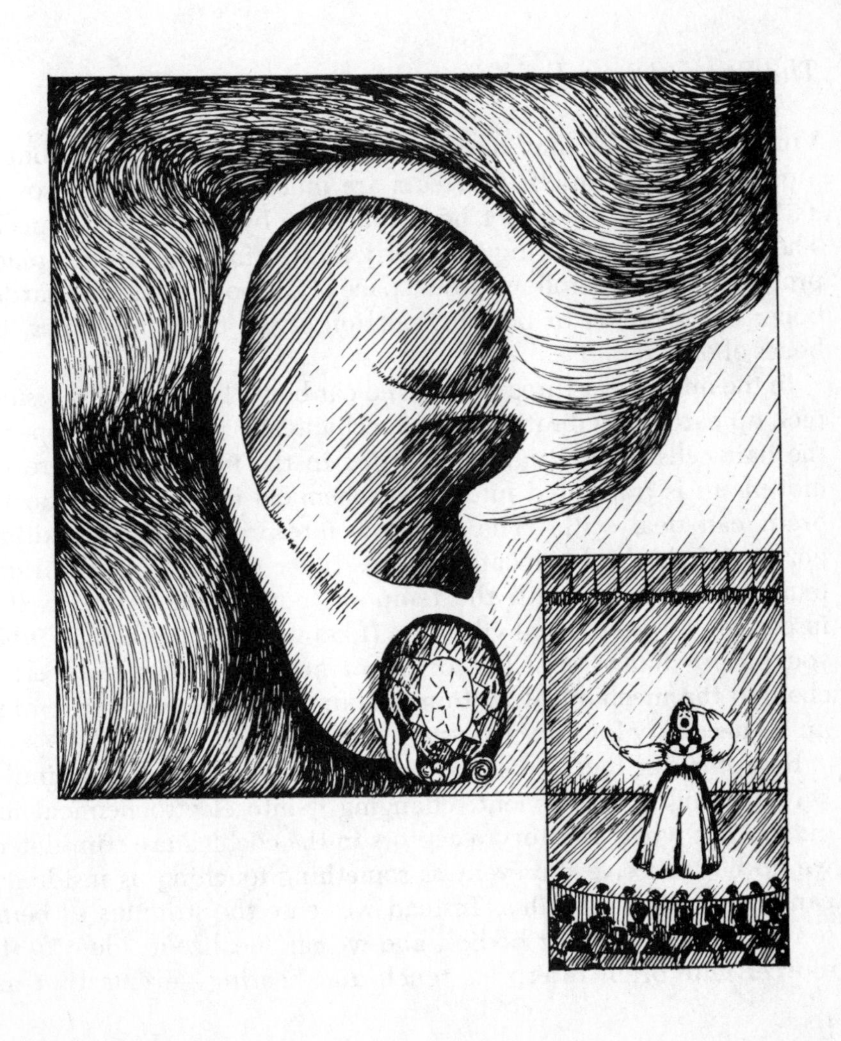

HEARING (formed from the verb "to hear," O. Eng. *hyran*, *heran*, a common Teutonic verb; cf. Ger *horen*, Dutch *hooren*; the O. Teut. form is seen in Goth. *hausjan*), in physiology, the function of the ear, and the general term for the sense or special sensation, the cause of which is an excitation of the auditory nerves by the vibrations of sonorous bodies.

The Ear and Sound

Vincent van Gogh may have cut off a pinna—an earflap—but he did not deafen himself. Our ears are much more than the sound-gathering scoops we direct here and there by swiveling our necks. The real organ of hearing is cleverly hidden in a much safer place, protected within a hollowed-out space of the body's densest, hardest bone: the rocklike, or petrous, portion of the temporal bones, the bones of our "temples."

In the middle ear three tiny linked ear bones known as the *ossicles* pick up eardrum vibrations and mechanically pass them deeper to the hair cells in the snaillike *cochlea* in the inner ear, where the movement is translated into electrochemical impulses, a form the brain can deal with. The sound is interpreted in the auditory portion of the brain, located on the inner surfaces of the brain's temporal lobes, beneath the temporal bones. The brain itself is insensitive to mechanical vibration (I know, it doesn't seem so when you have a hangover). The organ of hearing is a transducer: It changes the mechanical motion of sound into the electrochemical pulses necessary for nervous system transmission.

Receptors for touch, distributed over the body, work in a similar way, picking up movement, changing it into electrochemical impulses. But when auditory receptors in the cochlea are stimulated, we do not perceive the event as something touching us inside our ear, as a contact stimulus. Instead we sense the stimulus as being some distance from our bodies, and we can localize it, identify the source. Our brain interprets touch and hearing—events that are

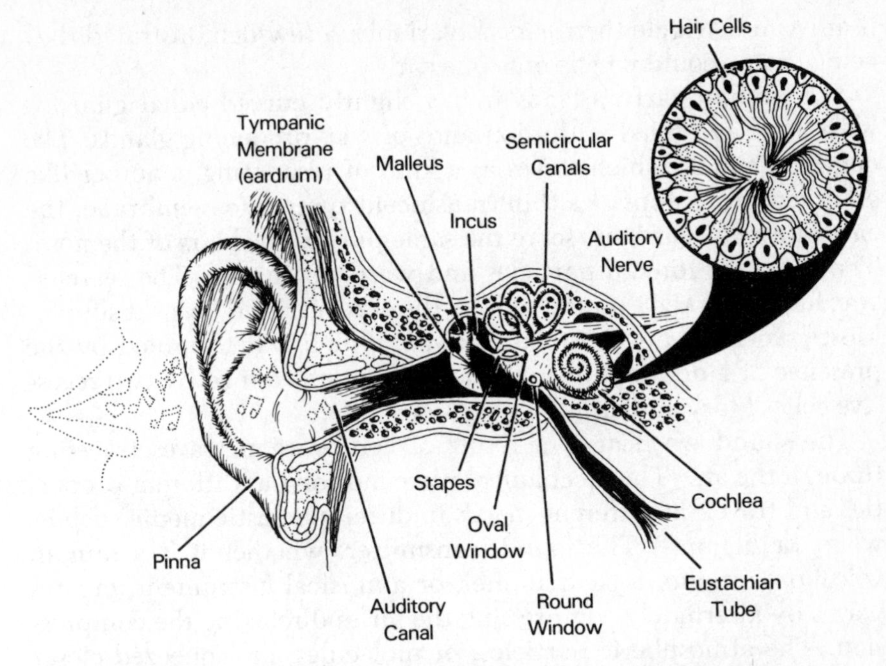

The ear is much more than meets the eye. The pinna is the sound-gathering funnel. The eardrum responds to sound waves and passes vibrations on to the ossicles, which rock in an air-filled cavity. The stapes rocks against the cochlea, setting the fluid in motion, and the appropriate hair cells respond. The auditory nerve carries the message to the brain, where interpretation of stimuli takes place.

basically the same at the receptor level—in two totally different ways.

The *pinnae* or external ears are characteristic of most mammals, although they developed much later in evolutionary time than did the hearing mechanism. The pinnae are also movable in most mammals—note how dogs and cats "perk up" their ears—and may serve a dual function as heat radiators. The desert-dwelling jackrabbit, for instance, has huge ears that are well supplied with blood vessels, through which it loses heat. The snowshoe hare, in contrast, has short ears to minimize heat loss. Some of the grazing animals have pinnae like ear trumpets that they can swivel in different directions. Humans move and cock their heads, instead, to use their pinnae. A few humans still have functional auricular muscles, but most have lost the power to move their ears. I liked to ask my biology students if any could wiggle their ears. Invariably, a few demonstrated they could. Some could wiggle only one ear.

The human earflap leads into a slightly curved canal guarded with hairs and lined with thousands of wax-producing glands. The one-inch canal, which serves as a kind of resonating chamber like an organ pipe, stops at a thin translucent tympanic membrane, the eardrum. The ear hairs serve the same purpose as hairs of the nose: They capture foreign particles and ward off insects. The secreted wax lubricates the skin that covers the eardrum and keeps it supple. Most people have wet, sticky earwax, which is determined by the presence of a dominant gene; the earwax determined by two recessive genes is dry and crumbly.

The sound we hear is a result of compression waves traveling through the air. (The mechanical wave motion has different properties and travels at different speeds in different elastic media such as water or helium.) The sound transmitter, whether it is a human voicebox, a radio, a jackhammer, or a musical instrument, creates waves by alternately compressing the air and relaxing the compression. The atmospheric particles, or molecules, are squeezed closer together than normal, then pulled farther apart than normal. The sound wave moves outward from the vibrating body, like ripples from a pebble thrown in a pond. The molecules themselves vibrate about an average resting place, like people jostling in a crowd. They

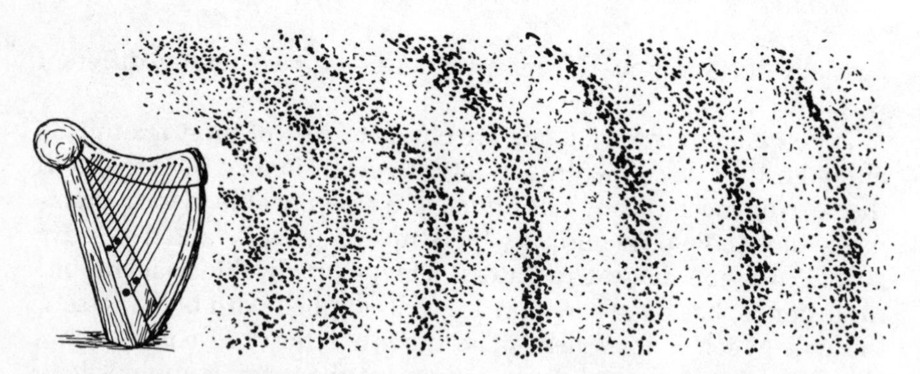

Sound waves are a result of alternate compression and rarefaction of molecules caused by a vibrating object. The speed at which the waves travel depends on the medium. Sound waves travel through air at approximately 340 meters per second; they travel even faster through water, and faster still through metals.

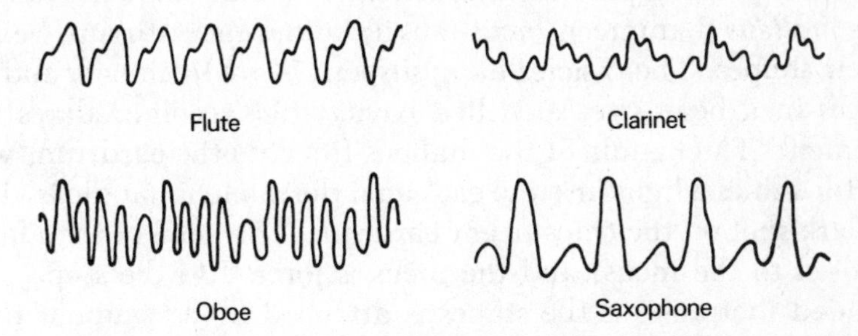

Sound waves made by musical instruments are regular in nature. A flute's high C can be distinguished from the same frequency produced by a violin because of the differences in the relative intensity of the harmonics in the two tones, and these differ because of the different methods by which the sounds are made. Complex sounds such as those produced by a symphony orchestra produce additive waves, but our brain is capable of sorting them for us. (From Physics of Musical Sounds, *© 1979, Van Nostrand Reinhold. Reprinted with permission from John Askill)*

do not advance with the wave. A sound wave weakens as it moves away from the source and may be reflected, or bent, by obstacles, so that sound reaching the ear may be different from that originally generated.

The soup-can telephones we made as children transmit sound through the longitudinal waves of a taut string (nylon works best; if the string is cotton, waxing it makes it stiffer and a better transmitter). We can also talk through a length of hose, which has the same effect as a doctor's stethoscope.

Sound-producing vibrations are repetitive and are measured in cycles per second (cps), or hertz (Hz), by acoustic engineers. The ears of children respond to frequencies as high as 40,000 Hz, but by adulthood the human sound spectrum has narrowed considerably, mostly by loss in the high-frequency range.

Musical instruments create variations in pressure with a regular pattern. The sound waves we translate as noise (unwanted sound) may have irregular patterns. Sounds created by the human voice as vowels and consonants have characteristics, respectively, of both regular and irregular sound pulses, of music and noise.

When airwaves strike the eardrum, the taut membrane vibrates at the same frequency as the oscillating waves. The vibration of the eardrum is picked up by the middle ear's ossicles, known separately as the *malleus* (hammer), *incus* (anvil), and *stapes* (stirrup) because of their shapes. The ossicles hang suspended by ligaments and tiny muscles in a bean-size, air-filled cavity that accommodates their movement. The handle of the malleus fits onto the eardrum; when a doctor shines a light into the ear canal the malleus handle is visible as a dark spot on the translucent eardrum. The head of the malleus is jointed to the incus, and the incus is jointed to the stapes. The expanded footplate of the stapes is attached by a ligament to the oval window of the fluid-filled cochlea of the inner ear, where it can rock in and out like a piston.

The ossicles are not just a Rube Goldberg device for getting from the outer to the inner ear. The middle ear's contribution to hearing is crucial. Fluid is more resistant to movement than air is. If sound waves were applied directly to the cochlea's oval window, they would not have enough pressure to move the fluid back and forth. The middle ear picks up the motion from the eardrum, which has an area 22 to 30 times that of the oval window, and increases the force with this ingenious lever system. The sound waves received by the eardrum are converted into more energetic vibrations that can be transmitted through the fluid of the inner ear's coiled cochlea.

The eustachian tube connects the air-filled cavity of the middle ear with the back of the nasal cavity. Through the eustachian tube, air pressure inside the eardrum is equalized with outside air pressure, important for proper sound conduction. The eustachian tube opens with swallowing, yawning, or nose blowing. When we fly on an airplane, or drive up or down a mountain road, we feel the air-pressure changes on our eardrums. By yawning, chewing gum, or blowing our noses, we can usually open the eustachian tubes and equalize the pressure in the middle ear. Tiny babies don't know this, and their crying from ear discomfort is usual as an airplane begins to ascend or descend. Scuba divers are taught to hold their noses and blow to equalize pressure in their middle ears as they descend.

The inner ear, known as the labyrinth (which means "a maze" or "something bewilderingly involved"), is fluid filled. Its hearing component is the snail-shaped cochlea, about the size of the fingernail on my little finger. The cochlea's compartmentalized interior is lined with more than 20,000 microscopic hairlike nerve cells arranged on a stretchy membrane. Each hair is tuned to a particular frequency or vibration. The hair cells at the base of the cochlea resonate at high frequency and those at the tip of the cochlea resonate at low frequency. Small animals have smaller, shorter cochleas than do large animals and generally communicate on higher-frequency bands.

When the stapes of the middle ear taps at the oval window of the cochlea, the cochlear fluid begins to vibrate. If middle C is sounded, for instance, the cochlea's middle C hair cells vibrate. The movement of these few hair cells sends a message to the auditory nerve cable, which has more than 30,000 transmission circuits leading to receiving cells in the auditory portion of the brain.

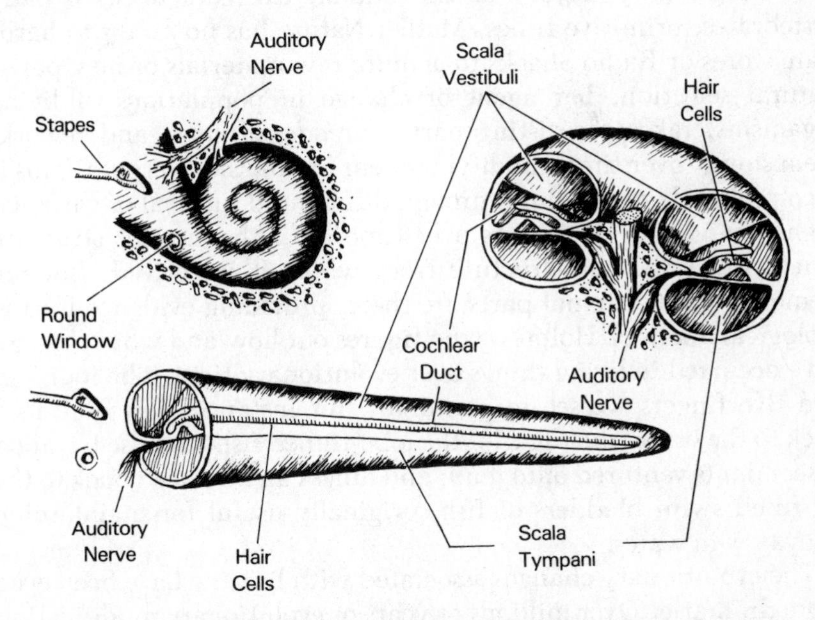

The cochlea of the inner ear is a fluid-filled spiral—an actual hollow in the skull. The pressure changes caused by vibration of the stapes produce the movements that stimulate the hair cells. When the cochlea is unrolled (possible only on paper), we see three compartments, two continuous with each other at the tip, the other separate, with a flexible membrane upon which the hair cells sit. The cochlear fluid is incompressible, so that when the stapes pushes the fluid at the oval window, the round window bulges.

Loudness is determined by intensity of movement—the more displacement back and forth, the greater the number of nerve impulses that are sent to the brain.

Closely associated with the cochlea are three loops of tubing, the semicircular canals, used in monitoring balance and equilibrium. They also contain fluid, and each loop is oriented in a different direction. They operate similarly to the cochlea, through hair cells sensitive to fluid displacement. One loop detects up-and-down motion, another detects forward motion, and a third detects lateral motion. The brain uses this information about fluid displacement to adjust body posture.

Jaws into Ears, or Installation of the Equipment

The evolutionary history of the human ear goes back to early vertebrates, primitive fishes. Mother Nature has no access to hardware stores or Radio Shacks to acquire raw materials or new parts. Natural selection, her agent of change in populations of living organisms, takes preexisting parts—*preadaptations*—and reworks them slowly over time, modifying them for another purpose. This is accomplished by selecting, through differential survival of variants, slight changes or deviations over time until the current structure barely resembles the structure from which it is derived. But the elements of the original parts are there, providing evidence for the biological Sherlock Holmes, who figures out how and what changes have occurred in living things over evolutionary time. The four legs and five fingers we see in mammals, for instance, can be traced back to the limbs and digits of the lobe-finned fishes whose tetrapod descendants ventured onto land, and lungs can be traced back to the air-filled swim bladders of fish, originally useful for maintaining buoyancy in water.

The evolutionary changes associated with hearing have been even more dramatic. Over millions of years of evolutionary modification a breathing structure was transformed into a feeding structure and finally into a hearing structure. The vertebrate hearing apparatus has its origin in the gill bars of early jawless fishes. The gill bars eventually became jaws, structures to help dismember and consume prey. The jaws in turn became reduced into the small ear bones in the premammalian reptile line, and other bones took over the function of jaw hinge.

Premammalian ears probably were first ears to the ground, functioning to detect substrate vibrations. Later they received low-frequency sounds transmitted through the air. The sense of hearing continued to elaborate (as, presumably, did vocalizations used in challenging and courting), the reptiles and birds evolving in one direction, the mammals in another. A few modern reptiles make good use of hearing, and some birds have clever hearing adaptations, as we shall see, but it is within the mammals that we find examples of hearing as the highest technology: the high-frequency echolocation of bats, the exquisite elaboration of the sense in whales and dolphins, and the uncanny sensitivity of cats. In our own group we find incredible discriminative capacity for one of the more complex waveforms—human language.

Several factors have contributed to the remarkable sensitivity of mammalian hearing. For millions of years, during the reign of the dinosaurs, the mammals were no larger than cats, and their small bony sound-conducting pathway presumably was very sensitive because of its size. The cochlea became coiled in the mammals (in reptiles and birds and in the primitive mammal, the platypus, the cochlea is straight), and with its greater length and thus greater surface area, came greater capacity for frequency range and discrimination. The early mammals certainly must have used sound to communicate with each other, but they also survived millions of years on Earth with meat-eating dinosaurs. They were probably under strong selective pressure to use their hearing to find prey (especially if they were sneakily active at night when vision was not as useful to this end) and to avoid being eaten. Elaboration of the hearing mechanism and the evolution of the brain's abilities to interpret the sound information it received went hand in hand.

Tree shrews, modern mammals much like the primitive ancestors of the primate group, have a very high frequency range of hearing. With the evolution of the primate group—bushbaby, monkey, chimpanzee, human—the high-frequency boundary collapses, from 60,000 Hz in tree shrews to 20,000 Hz in humans. Because the size of the ear structure does not directly correlate with this dramatic shift in frequency, the change in frequency boundary is most probably a result of selective pressures associated with improved within-species social communication.

We come from a long line of communicators.

The Human Voice and Music

Not long ago I witnessed a live performance by a supreme Italian musical instrument: Luciano Pavarotti. Pavarotti's voice overcame the symphony orchestra that accompanied him. His voice stimulated the mechanoreceptors of my skin as well as those of my ears. Was he real? About an hour later, he unexpectedly brushed past me in a nightclub, red scarf flowing. I levitated. Could it be, I thought, that he was in a private room, eating pasta and drinking wine, laughing, talking, using his voice like a mere mortal?

Yet art imitates nature. Nature is usually there first.

As a source of musical sounds, the human voice is the most versatile of the musical instruments in its possible variations of pitch (roughly corresponding to frequency), loudness, and the quality of sounds it can produce. The human voice is like a wind instrument, a stringed instrument (the vocal cords like strings), a percussion instrument (with resonating cavities), even like a bagpipe.

The vocal organs—lungs, windpipe, throat, larynx, nose, and mouth—are primarily for breathing and eating. Sound production is what they do secondarily. The lungs act as bellows. The vocal cords, or folds of the larynx, vibrate as double reeds. The cavities of the throat, nasal sinuses, and mouth act as resonating chambers, with the size of the mouth cavity varied by jaw movement.

The source of energy for the production of sounds is the stream of air coming from the lungs as we exhale. The maximum volume that can be exhaled is called the *vital capacity.* Vital capacity is usually about five liters. (Lung capacity is easy to measure in biology laboratory, with students exhaling into a spirometer or into a homemade gadget in which water is displaced and measured. My student with the most memorable vital capacity, about six liters, was a trumpet player, as I recall. Jimmy N., where are you now?) We can control exhalation, forcing air from the lungs up the windpipe to the voicebox, a distance of about 12 centimeters in mature males, 10 centimeters in females.

The vocal cords, two bands of skin at the top of the windpipe like a double reed clamped at both ends, are the source of sound in the voice. Closed, they shut off the airflow. Open, they allow air to pass through to the throat. When they are vibrating, they cause the air stream to be chopped into puffs, the frequency of which is the frequency of vibration of the vocal cords. The average length of the cords in a male is 2.4 centimeters; in a female, 1.5 centimeters. Their tension and their effective vibrating length are varied by the muscles attached to them—and these muscles can be trained, like the muscles of an athlete. Frequency of vibration is determined by the cords' length, mass, tension, and by the air pressure on them.

Male and female children's voices sound alike because there is little difference in the sizes of their vocal cords. The range of pitch increases with maturation. During puberty the size of the males' vocal folds increases, and adolescent boys have to relearn how to control the pitch of sounds they produce, which may take a year of intermittent squeaking. Their speaking and singing voices drop in pitch about an octave when their vocal cords more than double in size. Maturing female vocal cords increase in size only about 30 percent.

Low pitches roughly correspond to low frequencies, and high pitches to high frequencies, but pitch is a kind of psychological or subjective sound quality that enables comparison of one sound with another. We react to frequency, but what the brain actually decides it hears is pitch. About one person in a thousand has perfect pitch, which seems to be a hereditary trait. These people can name the pitch of a tone or a combination of tones or sing a given note without comparing it with any reference tone, provided that they were trained with an in-tune instrument. Others are not hopeless; musical training can help them.

The ranges of the human singing voice:

soprano	262–1047 Hz
alto	196–698 Hz
tenor	147–523 Hz
baritone	110–392 Hz
bass	82–294 Hz

These frequencies correspond to notes on the musical scale.

A good singer has a certain amount of vibrato and tremolo. Vibrato is the periodic change in the frequency of a note by 5 to 10 Hz at constant loudness, whereas tremolo is a periodic change in loudness at constant frequency.

As both men and women age, their pitch range begins to decrease as a result of loss of elasticity of the muscles in the cartilage of the larynx. Tenors become baritones; sopranos become mezzo-sopranos. Their speaking voices, like my aunt Fon's, sometimes acquire a tremolo. And Pavarotti retires.

We are superb at discriminating the complex waveform of the human speaking voice. A familiar voice answers the telephone with only a hello and we recognize it. Male speaking voices transmit at an average frequency of 125 Hz, females at an average frequency of 250 Hz, and telephone companies have studied how to most economically transmit the sound frequencies of voices. Telephone-voice transmission is much better today than it was when I was a child. Radios, which can transmit music, have wider transmission bands.

Sometimes our sensory wires get crossed. A crossover of interpreting sensual information occurs to some degree in all of us. In the extreme, this is known as *synesthesia*, and people in whom this occurs are *synesthetes*. Any of the sensations can be mixed, but the auditory-visual mix is the most common. Synesthesia has been recognized for a long time. More than 100 years ago, those with synesthetic ability were thought to be extremely creative, analyti-

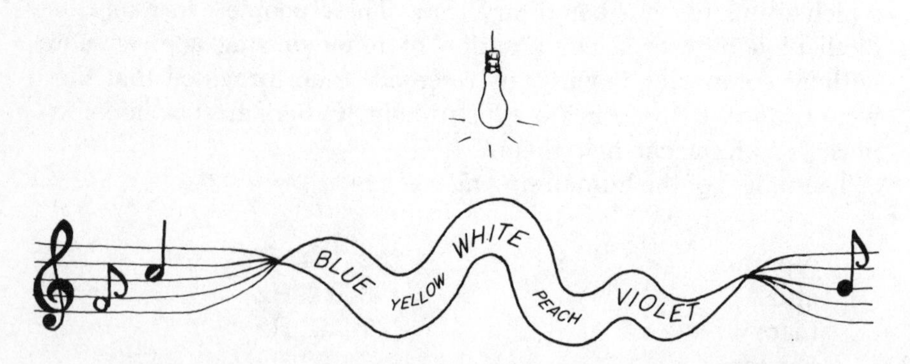

The senses do not operate separately from each other, but provide an integrated picture of our environment. In extreme crossover cases, known as synesthesia, stimulation of one sense will result in vivid perception in another sensory mode.

cal, and intelligent yet emotionally maladjusted. But they may just have some unusual interpretive wiring. A recent study of a large group of undergraduates at the University of Minnesota revealed that synesthetic abilities form a continuum. The students filled out a questionnaire about their synesthetic experience, then were given intelligence and personality tests. The synesthesia tests, in which they were asked to relate music with color and mood, were administered with a tape recorder.

Most students, male and female, associated sounds with imagery. In general, the students associated highest-pitched tones with the brightest colors, middle tones with middle colors, and lowest tones with darkest colors. They saw high tones (4000 Hz) as white, yellow, and pink; medium tones (1000 Hz) as blue and green; and low tones (200 Hz) as brown, gray, and black. In mood-color relationships, yellow was chosen to describe joyful music; black and gray, sad music; red, vigorous and exciting music; and blue, tender music.

Different Wavelengths

When I was a zoology student, one of my professors took us on a field trip in southern Arizona to listen for bats. Bats work the night shift, capturing insects. They are as diverse as birds, but humans are much more aware of birds and comfortable with them because we can relate better to birds' frequencies and timetables.

At night we sat around a fire near a streambed lined with tall cottonwoods. I knew this about insect-eating bats: They will fly over water or around streetlamps to capture insects attracted to the water or the light. Our fire died to coals and the first bats began to fly silently above us, fluttering delicately and swooping for insects. Then the professor turned on a receiver that translated the bats' inaudible sonar signals (25,000 to 100,000 Hz) into clicks. The previously quiet night was suddenly surprisingly noisy with bats. We had extended our peepholes, much like the seventeenth-century Dutchman Antoni van Leeuwenhoek had when he looked through the microscope lens he had made and saw a hitherto invisible world of cavorting "animalcules." They had always been there but out of sensory range.

Like radio stations, animals have their own communication bands, and our appreciation of animal communication has developed slowly because of the natural limitations in our own perceptual systems and sensitivities. Much as we realized that Earth is not at the center of the universe, our egocentric species has discovered that our nervous systems and what we sense in our environment do not constitute the whole picture. Whole noisy worlds of animals are communicating with each other just outside our tuning.

In the late eighteenth century, the Italian naturalist Lazzaro Spallanzani caught some bats in a bell tower, blinded them, and released them. When he later recaptured four of the bats, he found their stomachs were gorged with insects. They were obviously not relying on their eyes to catch insects. Could it be ears? Some bats certainly have big ears.

The bat scientists of this century have learned much more about the workings of bat senses and their system of echolocation for tracking prey. Donald Griffin and others have shown that bats use a sophisticated sonar system to hunt and avoid obstacles. Cruising bats emit four or five pulses of ultrasound every second. When the bats detect something, they step up the rate to as much as 200 pulses per second. The sounds bounce off the object and reflect back to the bat. Even though the echoes are much fainter than the sounds they emit, bats are very good at detecting their own noises bouncing back. Griffin let bats fly in a fruit-fly-filled room closely strung with thin piano wires. They avoided the wires and caught the flies. He also tried to jam their sonar by broadcasting broad-band frequencies. The bats were good at discriminating their own echoes. Griffin found, however, that beeping bats leaving or entering their cave would often crash into a net. He called this the "Andrea Doria" effect, after the Italian cruise ship that crashed into another ocean liner despite having the most advanced radar system of the time. Sometimes bats don't pay attention either.

As predators get better, prey get better at avoiding them. Natural selection again: survival of the fittest. Major prey items of insect-eating bats are the night-flying moths. Moths in several families have simple ears on their thorax that make it possible for them to hear the high-frequency sounds made by bats. If the moths detect the bats some distance away—and a moth can detect a bat at a greater range than the bat can detect it—they fly in the opposite direction. If the sounds indicate a bat is nearby, a moth takes evasive action, flying crazily, looping-the-loop, or dropping suddenly to the ground or to a bush.

Other insects listen to one another. Male fruit flies vibrate their wings when they court females, who detect the vibrations with their

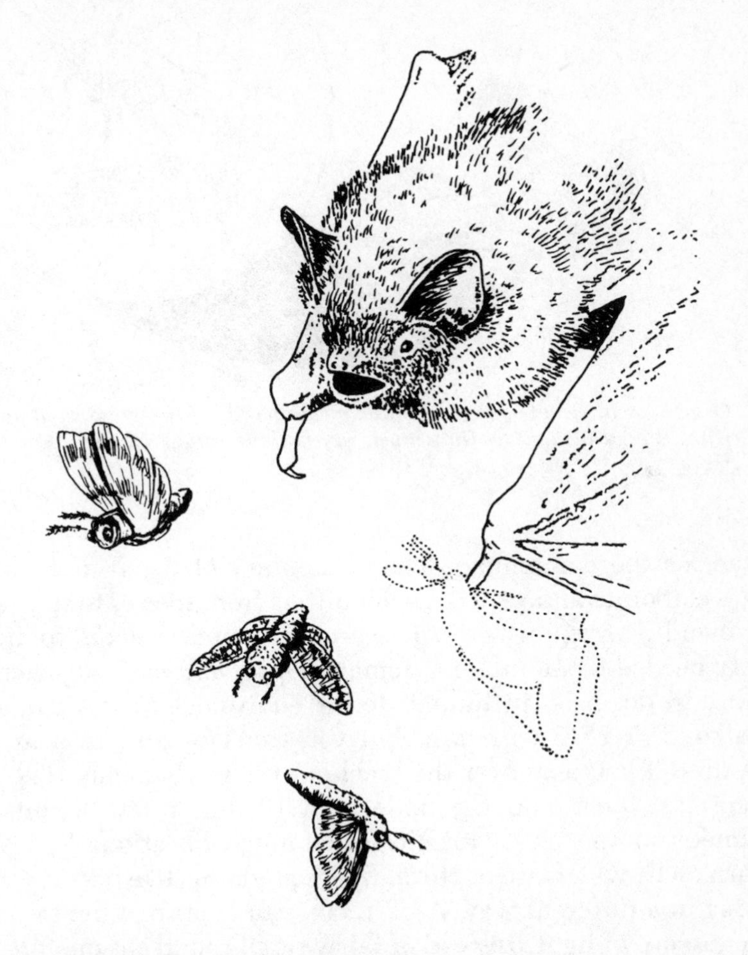

Bats use a sophisticated and lightweight natural sonar system to avoid obstacles and hunt night-flying insects. They emit ultrasound from 20,000 to 100,000 Hz and detect the faint echoes that bounce back. Moths in some families hear the bats with their thoracic ears and take evasive action. (From the illustration by John Langley Howard from "More About Bat Radar," by Donald R. Griffin. Copyright © 1958 by Scientific American, Inc. All rights reserved. Adapted with permission from Scientific American, Inc.)

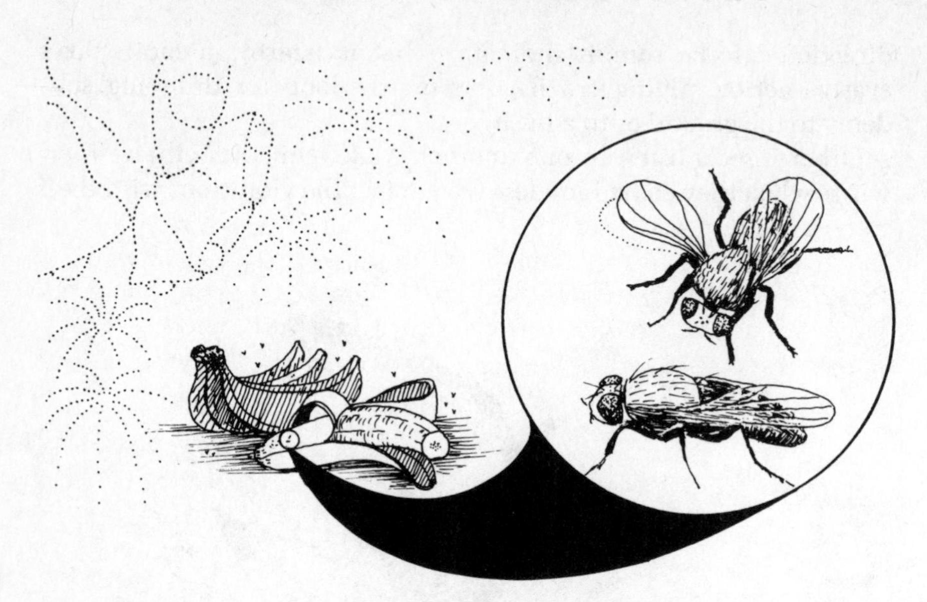

Over the fruit bowl or in the produce section of the supermarket, male fruit flies are making fine music with their wings as they court females.

aristae, or short antennae. The vocabulary of the fruit fly's love song is elaborate and varied, and it differs from species to species, so they avoid mating mistakes. Researchers found that the sound intensity needed to stimulate a female *Drosophila melanogaster* was about 115 decibels, in human terms equivalent to the climax of Tchaikovsky's *1812 Overture.* But you won't be able to hear fruit flies, even if you lean over the bowl of overripe bananas. The song doesn't last long, and the male fruit fly has to be within two millimeters of the female to deliver that much vibration.

Barn owls, which are noctural flying predators like bats, use their hearing in a different way. A barn owl can capture a mouse in the total absence of light. Instead of fairly small round ear openings, as most birds have, owls have long vertical slits, nearly as deep as the head itself. The facial disks characteristic of barn owls indicate their presence. The edges of the disks are fringed by short, stiff feathers and move to control the size of the ear opening, making owls able to scan for sounds like mammals do by moving their pinnae. The ear openings of the owl are also asymmetrical. The left opening is higher and probably exaggerates the effect of displacement of sound to one side or the other. The time difference is tiny, only 0.00003 seconds, but it seems sufficient to indicate where the source of the sound lies.

The owl also has a large inner ear and a larger auditory portion of the brain than other birds of similar size. The range of frequencies the owl detects is more limited than our own, but their maximum sensitivity region is in the high range, ideal for locating the high-pitched squeaks of rodents. The earflaps are best developed in owls of northern latitudes, perhaps because northern nights are longer and owls have to work harder to find those rodent dinners.

The owls' keen hearing is augmented by their ability to fly in nearly complete silence. No noisy flapping for an owl; its feathers are fringed for absolute quiet, and it glides to its prey.

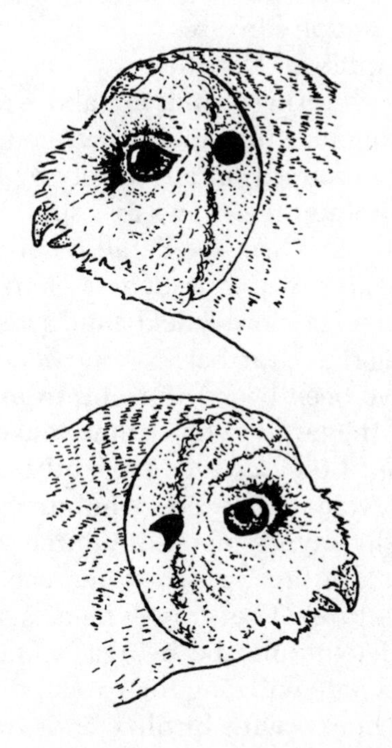

Barn owls have a variation on bilateral symmetry. Their asymmetrical ears help them to locate precisely, in total darkness, the padded footsteps of a mouse.

The cetaceans—whales and dolphins—use sound as their primary sense. Their eyes and noses are of little or no use in their watery habitat. They "see" with sound and speak, too, of course. Most have two kinds of voices, one for social communication, the other for navigation. Two kinds of sound receivers have been found in dolphins. One is in the jaw, through which sound travels in a thin oil to the eardrum, and one in an oil-filled "melon" in the forehead. Somehow the melon seems to broadcast and receive sounds simultaneously. The difference between the density of flesh and water is so slight that sound is not stopped by the surface of the body. It enters all over the head and body until it hits bone. Because the cetaceans (evolved from terrestrial ancestors) have received underwater sound through the vibrations of their whole skulls, external ears were unnecessary, and the whales gradually lost them. The ear channels were reduced to tiny pinholes and plugged with wax. Oil, a material less dense than water, channels sound to the inner ear. The sound picture a dolphin "sees" of a fish or a scuba diver probably would resemble an Xray to us.

Because their sonar hunting sounds are above the hearing range of fish, dolphins can exchange information without alarming the fish. Cetaceans can keep in sound contact simply by blowing bubbles as they exhale. They also pop their jaws and slap the water with their bodies. Like the bats, they can see detail by changing from low to high frequency. Because water is denser than air, sound disturbances compress and relax more quickly and sound zooms along five times faster, traveling much farther.

Whale sounds have been heard 15 miles from the nearest visible whale, and some of the larger whales may make sounds that travel hundreds of miles until they bounce back off the continental shelf. Whales may keep in voice contact over hundreds of miles of ocean. They are undoubtedly saying incredibly interesting things to each other. Dolphins speak in short phrases, and each has its individual whistle, like a signature. The gray whale is a Calvin Coolidge type—relatively quiet—while the beluga is known as the sea canary. A humpback whale will sing for 7 to 30 minutes, then repeat the song for several hours, with birdlike trills, arpeggios, and glissandos. The cetaceans represent another trove of information barely tapped by human peepholes.

Children, Language, and Ears

The first sound a human fetus hears, undoubtedly, is the soothing *lubb-dupp* of its mother's closing heart valves. After birth, mothers tend to hold babies over their hearts to calm them. In typical paintings of Madonna and child, the child is clinging to the left side—the heart side—of her breast. In his book *The Throwing Madonna*, William Calvin has suggested that this phenomenon plays a role in the development of the left brain/right brain dichotomy: Females held children with their left hand and threw rocks with their right.

Babies, he says, have also been shown to recognize voices they have heard before they were born (their mothers reading Dr. Seuss's *The Cat in the Hat*, even soap opera theme songs), and they prefer human speech to other sounds. Librarians have urged pregnant mothers to read to their developing fetuses, to give them early on a sense of the rhythms of language and a feeling for literature. Although some have claimed that a child's I.Q. can be raised by sending it to "prenatal school," where mothers-to-be talk to their bulging middles through megaphones, the claim is impossible to test. Intelligence capacity seems to be more than 50 percent genetic, but psychologists generally agree that an early enriched environment of stimulation has a positive effect. Much of that enriched environment is language. Children are programmed to listen and to begin the language process as soon as they are born. The template is there. A universal process of development has been identified: crying, babbling, echolalia (repeating), first word, one-word, two-word, three-word, then complete speech.

By the time a child is five or six years old, he or she has spent more than 20,000 hours listening to the world and listening to and learning to interpret the speech of those nearby. A child can recognize 10,000 complex sequences of syllables and relate them to events in his or her surroundings. If more than one language is used routinely at home, the number of recognized speech units is even higher. Other languages remain foreign because they lack the familiar sequences. Comparable lengths of listening and practice time are needed to add another language with full appreciation for tones of expression and nuances of meaning. Children who hear several languages early on sort them out quickly, however. One example is Sofia, the four-year-old daughter of my friend Monique, a French-Canadian, and her husband Jesus, a Mexican. Without confusion, Sofia speaks French with her mother, Spanish with her father, and English with others. Sofia has an enriched language environment.

This channel for learning about the environment is closed for a child who is congenitally deaf. The worlds of the prelanguage development (prelingually) and postlanguage development (postlingually) deaf are strikingly different.

Tom Clark, whose father was deafened at age ten by the high fever of spinal meningitis, works today in a national program aimed at identifying deaf babies and working with home-based programs that will help the children as much as possible to develop language early on—to make the home a meaningful place for communication and language. Language is a natural phenomenon that is learned in the home, and without language, education is difficult. "If a child has language early, it's on its way," Tom says.

Tom's father had a full set of language inside him when he was deafened, Tom says. But he could no longer attend public school, so he attended the Utah School for the Deaf. Later he graduated with honors from Gallaudet College, took civil engineering courses at the Massachusetts Institute of Technology, and became supervisor of the Kaibab National Forest in southern Utah. Another boy in his town, born deaf, never learned to talk and never developed speech.

On the surface the loss is the same, but the difference is that linguistic individuals still have the developed language in their minds. At age three and a half or four, Tom says, language is rooted. If a child is deafened at younger than 30 months, language has not taken its hold, and whatever is present disappears quickly. In general, a deaf child's learning grows in proportion to a hearing child's at a ratio of about two months to the year. The program Tom works with is designed to accelerate that growth by intervention at critical developmental periods.

A partial hearing loss also shuts out sounds of everyday existence from which children learn. Minor undetected hearing losses have been linked to learning disabilities. Many may result from middle ear infections—*otitis media*—which are common afflictions of children, second only to the common cold. In children, the eustachian tube is shorter and less angled than it is in adults, so bacteria have a shorter and easier route to the air-filled cavity of the middle ear. Recent studies indicate that children of European ancestry may have an inherited tendency or susceptibility to the infection.

When the middle ear air pocket fills with the fluid of infection, the ossicles are unable to rock properly. Children may have frequent earaches, or the infection may be discovered when a child is treated for another problem, significantly often a speech problem. Because children with otitis media cannot hear properly, they are unable to mimic voices in their environment. Continued infection can cause permanent damage, but the infection is usually treated successfully by inserting a tiny drainage tube into the child's eardrum, a medical practice with a 100-year history. The tube stays in place from 8 to 14 months before being extruded into the ear canal by the child's growing body. After the tube pops out, the eardrum hole closes naturally.

Audiograms

The ears of children respond to very high frequency sounds. But by adulthood, the human sound window has narrowed to about 15-15,000 Hz—a range of nearly ten octaves, with each octave representing a doubling of the vibration rate. Measurements of the hearing of men in their forties show that the upper limit of the ear's sensitivity falls about 80 Hz every six months.

Presbycusis, the loss of hearing as a result of aging, begins in adolescence but doesn't become noticeable until years later, when a person can't hear high-frequency sounds such as a ticking watch or certain high-frequency consonant sounds.

Loss of hearing in the high frequencies may explain why we cling to middle C, at 256 Hz, as a landmark in the musical scale, although it is well below the middle of the audible range of youthful ears.

Around 3,000 to 3,500 Hz, between F sharp and G in the fourth octave above middle C, humans are most sensitive to sound. This is related to the resonant wavelength of the auditory canal, which amplifies sounds in this frequency range. Near this pitch, a standing

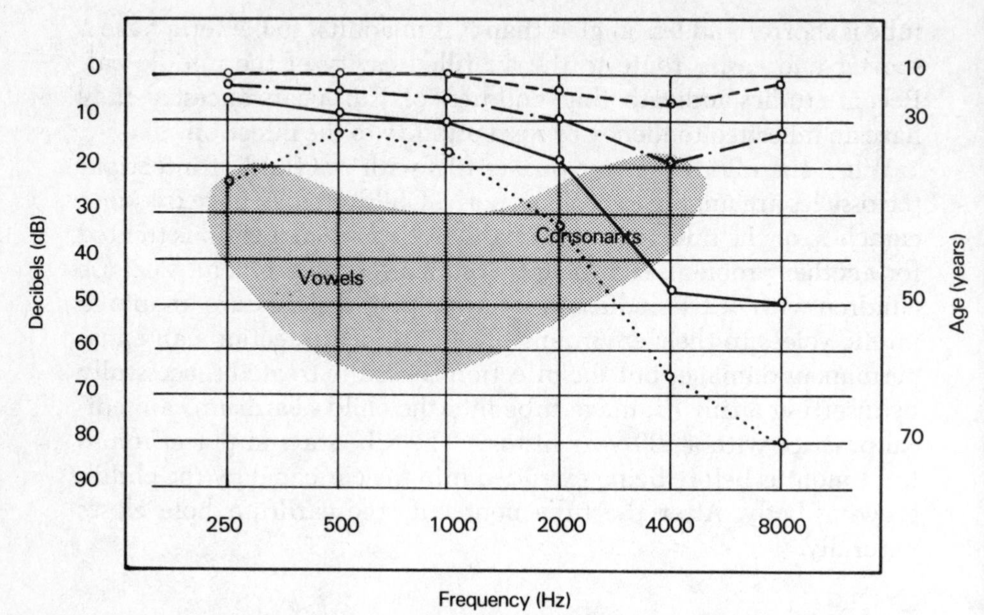

Presbycusis refers to the gradual loss of hearing, especially in the high frequencies, that occurs with age. First to be lost are some of the consonants essential for understanding human speech.

wave can be formed in the ear canal and requires the least energy for sound to stimulate the ear and give us a sensation. This is the frequency of piercing screams or alarm calls.

From more than an octave below middle C up to about 3,000 Hz, human hearing and a cat's are about equal. But the cat's ears respond far more sensitively than those of humans to sounds of higher pitch. At 13,000 Hz, in the highest octave an ordinary adult human ear will detect, a cat needs only a thousandth as much energy to identify the presence of vibrations. A cat has at least 40,000 fibers in the auditory nerve, compared with our 30,000. Its upper limit seems to be around 50,000 Hz. My cat hears me coming and meets me at the door, and yet a cat's hearing may not permit it to detect the calls of mice, which are pitched even higher. One mouse can hear the warning scream of another up to 95,000 Hz. (Cats are subject to hereditary deafness, as are humans. The genetics is a little complicated, but white cats and blue-eyed cats are more likely to be deaf than colored cats with yellow eyes. Undoubtedly, deaf cats owe much of their survival to loving human caretakers.)

The greater the amplitude of motion, the higher the pitch. The greater the force of vibration, the louder the sound. While frequency is measured in Hz, relative sound intensity (how much pressure sound exerts on a surface) is measured in decibels (dB). The decibel scale is humanly egocentric. Zero dB is the level at which an average young adult can barely detect sound; between 115 and 130 dB is the threshold of feeling and pain. The decibel scale is also tricky. On the decibel scale, an increase in one dB means an increase in intensity of 1.26 times, so that 20 points on the scale means a hundredfold increase in intensity. A sound of 100 dB is 100 times stronger than a sound of 80 dB.

Some approximate examples are given in the following list:

Sound	Intensity (dB)
absolute hearing threshold (for 1000 Hz tone)	0
streamflow, rustling leaves	15
soft whisper	20–30
quiet room	30
conversation	45–60
normal office level	50
crowd noise, city noise	60
vacuum cleaner	75
inside noisy car	80
electric shaver	85
heavy equipment	90
jackhammer, subway train	100
gas-powered lawnmower	105
loud thunder, rock band	120
pain threshold	130
shotgun, jet plane takeoff from 80 feet	140
rocket launch from 150 feet	180

The ossicles have minute muscles attached to them that contract when we are exposed to intense sounds. This contraction, known as the middle ear reflex, reduces the transmission of sound through the middle ear, particularly at low frequencies, and may help to prevent damage to the delicate cochlea. The reflex is too slow to provide any protection against impulsive sounds such as gunshots or hammer blows, but it helps to reduce self-generated sounds and is activated just before vocalization. It may also reduce masking of higher-frequency noises by lower-frequency ones.

The cocktail party effect is a familiar phenomenon. The average intensity level gets higher and higher, and people speak louder and louder, filtering out the sounds of other conversations. Soon the people speaking with each other must be within six inches of each other's mouths to hear their conversation and find themselves shouting.

For most of our evolutionary history, the rumbles of thunder or the roars of lions were probably the loudest sounds around. Beginning with the first metalworking, the human environment has become progressively noisier. Industrial hearing loss has been concomitant. Several hundred years ago metalworkers were expected to become hard of hearing and, if they grew old at the work, completely deaf.

Today, about 8 million Americans are exposed to on-the-job noise levels that may permanently damage their hearing. Audiologists can identify a noise notch in the audiograms that indicate hearing loss resulting from exposure at a certain frequency level. The notch may be temporary, perceived as a muffling of sound. It may be accompanied by tinnitus (head noises) and, at work, usually occurs during the first two hours of continuous exposure. Repeated exposure can cause permanent hearing loss.

Temporary deafness, however, has also been caused by "natural" sounds. Not long ago a young mother complained of temporary deafness after a screaming bout by her 11-month-old baby. Her physician husband measured the sound intensity of the baby's screams at 117 dB.

An audiologist who tests and diagnoses hearing loss in students, industrial workers, and people in hospital settings tells me she has identified hearing loss in young people, both males and females, that can't be accounted for by aging. At one high school recently, she referred for further testing 5 of 60 students she examined—that's 12 percent.

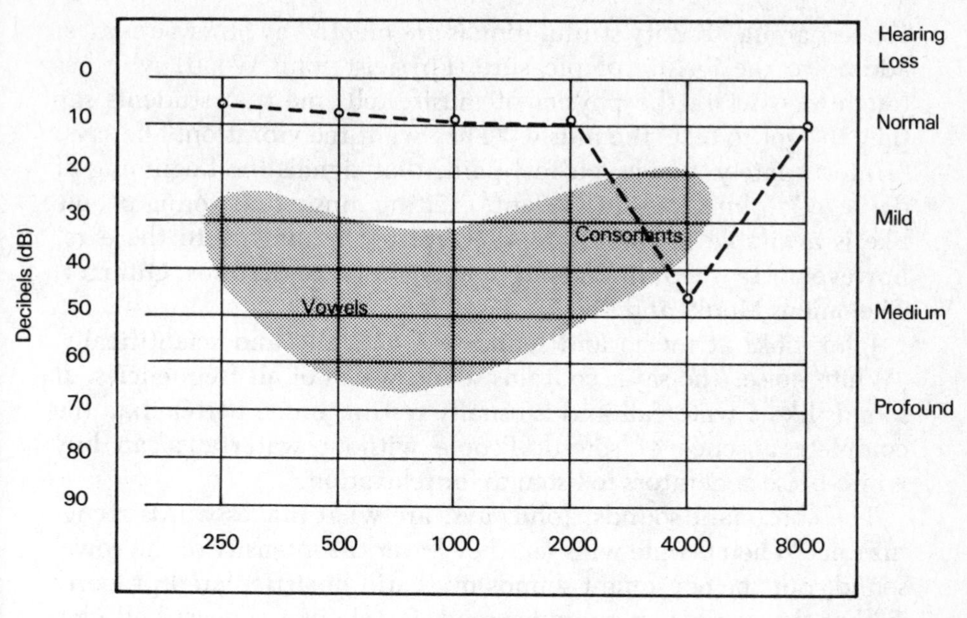

Audiologists can identify a characteristic noise notch that indicates exposure to loud sound in a certain frequency. The notch may be temporary or permanent.

I was in fear for my ears when, as a teacher, I chaperoned high school dances. The music was amplified to the point of pain. I know now that my middle ear ossicles were wiggling sideways. I was required to stand near the band as a kind of crowd control measure. I didn't have ear plugs. Subsequently I volunteered to take tickets at the football games to fulfill my extracurricular service expectations.

No one else seemed bothered at the dances. For them the loud music was fun. I have had students tell me, smilingly, of their ears ringing for hours after a rock concert. That seemed to be a measure of its success. Michael Koss, president of Koss Stereo Headphone Corporation, says the excitement that people, especially teenagers, get from high-decibel music results from activation of the peripheral nervous system by low frequency sound waves beating against the body. Koss contends that people can get "high" from this feeling, because it switches on the body's fight or flight mechanism, bringing a rush of adrenalin (a reason for battle music?). In a concert arena, sensory stimulation is augmented by other activities, adding to the feeling of pleasure. Physicist John Wood, who has taught courses in the physics of music, tells me that students say they've "got to feel" the music. "They want the vibration," he says. "Unfortunately, when you feel pain, that signals the beginning of damage." John says that vibrant, exciting music that young people like is available at intensities that are not damaging to the ears, however. He recommends the music of the Kronos Quartet, Thelonious Monk, and Philip Glass.

John looks at music and sounds aesthetically and scientifically. "White noise," he says, contains sound waves of all frequencies. It sounds like a waterfall and is usually restful, much better than the complete absence of sound. People without waterfalls can buy white-noise generators to listen to for relaxation.

The consonant sounds, John says, are what makes words recognizable. When people whisper, they lower the intensity of the vowel sound, but the consonant sounds must still be articulated properly so that the words can be understood. People in a concert hall may think their whispering can't be heard, but it's intelligible 15 rows away because of the consonant sounds. Cut out the vowels when you write words, he says, and you can still read them. That's what the Hebrews did on the Dead Sea Scrolls (remember JHWH?). Cut out your vowels and you can still be understood. I quit whispering in concert halls after I talked with John. *Shhh*, by the way, is a high-frequency sound, easily heard.

John tells me of his friend, another scientist, who had a hearing test to make sure he could hear the high-frequency bandwidths of an expensive new stereo set he was considering buying. The greater the bandwidth, the more of the musical harmonics, and thus richness, are transmitted. The brain puts in some of its own harmonics too.

Musical technology, like telephone transmission of voices, has gotten better and better. One of the reasons compact discs are so good, besides their durability, John says, is because they are digital rather than analog recordings. The digits are translated back into sound by the CD player. CDs don't cut off the tops of the sound waves and can capture more dynamic variation than can magnetic tapes or records, in which the recording process physically limits extremes in loudness. CDs always sound like the original transmission, he says.

We love to listen to recorded music. I have my own portable cassette player with headphones, which I find especially useful on airplane flights to mask the noise of the airplane motors. My friends wear them while gardening, mowing the lawn, walking, bicycling, jogging, shopping, studying, or writing. I see them everywhere. But their improper use may increase hearing loss.

In 1980 an ear, nose, and throat specialist at Hokushin General Hospital in Japan surveyed the hearing ability of 4,500 students from six high schools. He found 29 students had hearing difficulties in the high-frequency range. Of the 29, 21 were headphone addicts. They told him they listened to stereo headphones more than 24 hours a week at volumes averaging 87.9 dB.

Another study conducted on the streets of New York City found that much "ambulatory" music—music carried along with the listeners—is played at loudness levels upward of 100 dB, with some as high as 120 dB. Another study showed 30 percent of discotheque disk jockeys suffer hearing damage.

The Better Hearing Institute in Washington, D.C., has warned that the danger from earphones is greater when the decibel level is too high and exposure is prolonged. The U.S. Occupational Safety and Health Administration indicates that exposure to 95 dB and above for four hours or more a day is likely to cause permanent hearing impairment.

Hearing Loss

Partial hearing loss and deafness are widespread. In the United States, 14 million have hearing problems and 2 million more are totally deaf. Hearing loss affects one person in ten of the general population, and one person in three over age 60. As the hearing window shrinks with age, men are disproportionately affected.

Like other physiological happenings, such as digestion of food, clotting of blood, or the response of immune system cells, proper hearing depends on a cascading series of events. Hearing loss can be related to any of the parts of the hearing mechanism or interpretive parts of the brain.

A *conductive* hearing loss involves damage to the mechanisms that transmit sound energy to the cochlea and may be caused by something as simple as impacted earwax (or as exotic as a two-year-old squirting Super Glue into a sleeping father's ear, as reported in an American Medical Association newsletter), which prevents the eardrum from vibrating. A conductive loss can also result from fluid in the middle ear, which causes the eardrum to sink inward and stops vibration of the ossicles. Aging of the hearing apparatus may contribute in several ways to diminished hearing. For instance, as the eardrum ages, it becomes less supple and does not vibrate as well as a young eardrum. The joints between the ossicles can harden with age, impeding conduction. If middle ear infection is chronic, scar tissue can form and immobilize the ossicles. In *otosclerosis*, the stapes is frozen by bone deposited around the oval window of the cochlea, a condition that may be corrected surgically.

Old violinists have been known to compensate for conduction problems of aging eardrums and ear-bone linkages by touching their teeth to their vibrating violins when tuning them. In this way, the eardrum and bone linkages are bypassed, and the vibrating skull conducts high notes to the cochlea so they can be heard clearly.

The cause of *nerve deafness*, on the other hand, is seated in the labyrinth of the inner ear or in the auditory nerve. Two-thirds of the profoundly deaf have nerve deafness. A bout of spinal meningitis carries with it a high risk of deafness. The high fever of spinal meningitis burns out the auditory nerve, Tom Clark tells me. Thirty of the two hundred children in his program in Utah were deafened by spinal meningitis.

Exposure to a number of drugs or chemicals can cause temporary or permanent hearing loss, usually affecting metabolism in the cochlea or hair cells, or causing physical damage. Aspirin or certain antibiotics can cause ringing in the ears, for instance. Babies and young children are at especially high risk for autotoxic drug effects.

Trauma, cardiovascular disease, and allergies may contribute to *labyrinthine disease*, which can cause the ringing in the ears known as tinnitus, hallucination of movement known as vertigo, nausea, vomiting, and visual hallucinations, as well as deafness.

Menière's disease is related to excess fluid in the inner ear; a person with Menière's may experience years of fluctuating hearing loss, vertigo attacks, and tinnitus, with an end result of deafness. One of its victims may have been Joan of Arc (1411–1431). Joan heard voices and saw visions of saints who commanded her to go to the dauphin of France to persuade him to be crowned king and dedicate his kingdom to God. Joan was "strange," but she was not mad, writes the medical historian F. F. Cartwright. At her trial, Joan said that the voices first came to her when she was 13. At that time she could not understand what they said. Later the figures of saints appeared, she said, in the form of minute things, in great numbers, accompanied by bright light and a pleasant odor. Cartwright suggests that Joan suffered from intermittent attacks of severe tinnitus. The tinnitus was on her right side only, with bright light and dancing specks. The specks are symptoms of nausea, and Joan sometimes fell to her knees and vomited. In jail, her vomiting was attributed to the spoiled fish she had eaten.

The religious reformer Martin Luther (1483–1546) is known to have suffered from middle ear problems but with different apparitions: He thought the Devil was yelling at him.

(Hearing internal voices, however, is not so unusual. A survey revealed that one in ten people have heard voices. The schizophrenics—about 1 percent of the population, in whom auditory hallucinations are common—talk about them. The perfectly sane people don't usually discuss their inner voices at parties. For a treatise speculating on the adaptive significance of unexplained internal voices, and the sides of the brain talking to each other, see *The Origin of Consciousness in the Breakdown of the Bicameral Mind*, by Julian Jaynes.)

Ringing in the ears usually indicates ear damage. The 36 million Americans afflicted with tinnitus experience ringing, whistling, buzzing, roaring, hissing, and screeching sounds that may range from mildly annoying to incapacitating. Tinnitus may be constant over a long period, or it may occur after the ears have been subjected to sustained loud noise. If a doctor can detect the ringing, too, it is usually caused by an open eustachian tube, a blood vessel abnormality, or palate muscle contractions, all of which can be treated medically or surgically. Usually, however, the noise is heard only by the person experiencing it, and the causes are various, sometimes mysterious. Tinnitus caused by aspirin, antibiotics, or infection can be relieved when the drugs are discontinued or the infection cured. Maskers—electronic aids that substitute an interfering sound—and biofeedback can help alleviate the stress caused by tinnitus.

More than 70 kinds of hereditary deafness have been identified.

Some are congenital and severe. Some are associated with other kinds of diseases or malformations. Others are progressive. Even if all parts of the ear are intact and functional, damage to the auditory portion of the brain can affect hearing. If the interpretive part of the brain is impaired, hearing is also impaired. A person with *auditory agnosia* hears noises, but sounds have no meaning because of a processing fault in the brain.

The Noblest Faculty

In 1800 the composer Ludwig van Beethoven, then 31 years old, wrote a friend that his "noblest faculty"—his sense of hearing—had greatly deteriorated. He said his ears whistled and buzzed continually and that he avoided all social gatherings. "How is it possible," he wrote, "for me to tell people, 'I am deaf'?" Beethoven said he could not hear the high notes of instruments and singers at the theater. Low tones of conversation were audible, but he could not make out the words. "It is strange that in conversation people do not notice my lack of hearing," he wrote, "but they seem to attribute my behavior to my absence of mind."

Beethoven was aware of his affliction, but he was also obsessively concerned with protecting his employment, and thus his source of income. He feared that his "enemies" might learn of his trouble, and he believed every rival was an enemy. He thought that others were talking about him. To hide his handicap, Beethoven pretended he was absent-minded.

Beethoven was totally deaf by the time he was 47. Various hypotheses have been advanced to explain his deafness: typhus, venereal disease, even a violent tantrum (he was not a subdued man). He tried every remedy: ear trumpets, baths, ointments, rest. Nothing helped. He poured water over his head. He cut the legs off his piano and lay on the floor to play closer to the vibrations. He continued to compose, using an out-of-tune piano with broken strings.

Once in Vienna during his last years, he was mistaken for a vagrant and arrested late at night. He was wandering aimlessly, humming to himself, waving his arms and marking time with his feet. He had confided to a vicar that his deafness caused him the least trouble in playing and composing—he had the magic of his music rooted in his mind—but most in his association with others.

Recently, doctors at Columbia-Presbyterian Medical Center suggested that Beethoven's deafness was caused by *Paget's disease*, in which abnormal bone growth crushed his auditory nerves. His big head and massive brow are characteristic of Paget's disease, a disease that was undescribed at the time of his death.

Attitudes toward those with hearing impairments may have changed little since Beethoven's time. When I taught biology at Pima Community College in Tucson, several deaf students took my course. Helene, a hearing student, interpreted for some of them. Helene has bridged the deaf and hearing worlds. She is an only child of congenitally deaf parents, and she spent much of her childhood in a deaf environment. Signs were her first language. She remembers a time before she merged sign language and spoken language, and when she had the insight to combine the whiskers sign for *cat* with the spoken word *cat*. When she went to a speaking preschool and was asked her name, she finger-spelled it.

Helene told me that she still sees widespread intolerance of the hearing impaired, although society has made strides in educating the general public about hearing handicaps. "Deafness is an invisible handicap," she said. "If you see someone with a white cane, you have time to think about the handicap, but when you speak to someone and that person doesn't understand, most people don't have the patience to work on communicating. The handicap is lack of communication. The world is not made for people who don't hear."

Hearing loss also affects the development of speech, and so the handicap is compounded. Deaf people have been known as "deaf and dumb," or "deaf mutes."

An ear surgeon I spoke with agreed that people in general are impatient with persons who ask "What?" The problem has another dimension, he said: "There is no stigma attached to wearing glasses, but there is a stigma attached to wearing a hearing aid." A hearing aid signals "elderly" to many people. In our youth-oriented society, vanity often keeps people from using a hearing aid, when it would make life easier for them, as well as for their families and friends. Ronald Reagan, who has been known to cup his ears and ask reporters to repeat questions, may have some vanity problems about his own hearing, but in recent years he has undoubtedly helped the cause of the hearing impaired by his example. He appeared in public in September 1983, at age 73, with a hearing aid and later began wearing hearing aids in both ears. Reagan attributed his hearing loss to a 1939 filming incident in which he was playing the role of Secret Service agent Brass Bancroft and a gun was unintentionally fired near his ear. (His then press secretary Larry Speakes said testily, "What's age got to do with hearing?") Reagan said a doctor told him that "those little nerves were damaged, and they lost their sensitivity as time went on."

Reagan's 1983 appearance with the hearing aid was to announce a literacy campaign, but his hearing aid became the news. The nation's major newspapers prominently carried the story. Editorials and features followed. A *New York Times* editorial acknowledged that "in a country that worships youth and fitness, age and its inconveniences are often treated like dirty little secrets . . . whether age or accident caused the hearing loss is irrelevant: What matters is whether or not it can be corrected —and it can."

Problems of Detection and Paranoia

Hearing loss in all its dimensions and gradations is a challenging problem for professionals, as well as for those affected by it. Hearing loss is not a disease in itself, is not accompanied by pain, and is not always readily detected. It is often symptomatic of other problems and, even when it is not severe, can adversely affect personality. Several experts have told me that hearing loss can be more disabling than blindness, precisely for the reason Helene gave: It profoundly affects communication, and communication is essential for social interaction.

But physicians are generally emphatic that every patient with a hearing disorder can be helped in some way. Although a patient's hearing may not be improved, a patient can be taught to use whatever hearing is present, and attitudes and communication can be improved.

First, however, a victim must recognize the need for medical help. Because hearing loss most often occurs gradually and invisibly, the loss can go undetected and unrecognized by the victims and even by their close friends and family. Victims seem not to notice hearing loss until their hearing has deteriorated markedly; this is true of children as well as adults. A student who is doing poorly in school may be missing auditory information and may not realize it. Sometimes hard-of-hearing patients have been under a psychiatrist's care for a long time before being referred to an ear specialist for a hearing evaluation. Or they may have been nagged into getting a hearing examination by a spouse, friend, or co-worker who has been frustrated by communication problems. A classic kind of marital strife can be caused by hearing problems: A wife says her husband doesn't pay attention to her; the husband doesn't think anything is wrong.

Studies have shown that some elderly people described as paranoid—suffering from delusions of persecution—have those symptoms because they have been slowly losing their hearing and are not aware of it. In experiments in which student volunteers were rendered partially and temporarily deaf by hypnosis (they were told—those sneaky psychologists—only that they were involved in a study examining the role of hypnotism in creative problem-solving), the volunteers suffered symptoms of paranoia that are common in the elderly. The students were perceived by observers as becoming "more irritated, agitated, hostile and unfriendly" and were also measured as paranoid on tests of personality.

This experiment illustrated the potential role of hearing loss, as distinct from the vague concept of old age, in the development of paranoia. People of any age who are partially deaf without realizing it, said the experimenter, risk developing paranoid thinking, which is a result of "distorted perception of reality." As time goes on, a victim concocts elaborate rationale to make sense of the distortions.

Fluent on Their Fingers

In 1987 the deaf actress Marlee Matlin received an Academy Award for her performance in *Children of a Lesser God*, a movie about a relationship between a deaf woman and a teacher at a school for the deaf. The movie, which had been a successful play, was about love and communication—communication in a kind of language foreign to most moviegoers: sign language. William Hurt translated gracefully for watchers as he and Marlee signed.

Sign language traces its recorded history to Italian Benedictine monks and is a testimony to the human need to communicate. About A.D. 530 the monks, who had taken vows of silence, created a form of sign language so they could "talk" with one another. The sign language was passed down through the centuries, and monks many hundreds of years later used sign language to teach deaf pupils.

In 1775, when the Abbe de l'Epée started his school for the deaf in Paris, he had learned sign language from deaf people, modified it to approximate spoken French, and used it to instruct his students. Thomas H. Gallaudet was introduced to signs used in de L'Epée's school and brought the French sign language to America in 1816 when he and Laurent Clerc founded the Gallaudet School in Hartford, Connecticut. Sixty percent of the signs in American Sign Language are estimated to be of French origin.

Teaching fads and philosophies come and go in most areas, and instruction of the deaf has also seen major controversy and upheaval. The war of methods, a controversy over the best way to educate deaf children, arose in the eighteenth century and raged into the twentieth. Sign language had fallen into disfavor and was blamed for deaf children's lack of speech and for their poor grasp of spoken language. Sign language was accused of promoting clannishness among the deaf and was regarded as an outward sign of their infirmity. Deaf children look normal, some reasoned, so they should be taught to speak normally. It's not an easy accomplishment.

At Gallaudet College sign language was still taught formally, but the oral method was used in other schools. In oralism, children often had their hands tied or were forced to sit on their hands while they were instructed to speak in front of mirrors, using flashcards and headphones that might stimulate any residual hearing. Oralism and suppression of sign language succeeded only in driving sign language out of sight. Those compelled to use oralism at school used the more rapid sign language in other situations. One scientist, W. G. Noble, has stated that people born deaf will develop signing, if permitted, as a natural language. Signing, he writes, has now achieved the status in hearing culture of a language comparable with any other. It has always had that status, although sometimes clandestinely, in deaf culture. Today deaf people are involved increasingly in "total communication"; that is, they sign, finger-spell, and speak or mouth words simultaneously. Their communication involves all effective methods for transmitting and receiving information.

Samuel F. B. Morse (1791–1872), the inventor of the Morse code, and his deaf wife developed their own method of communication. They tapped code into each other's hands.

Alexander Graham Bell (1847–1922), a professor of vocal physiology, dedicated teacher of the deaf, and inventor of the telephone, was married to Mabel Hubbard, who had been deafened at age five by scarlet fever. Bell's mother's hearing was also severely impaired, although she was a good pianist, achieving feedback by fastening an ear tube to her ear and resting the mouthpiece on the soundboard. Bell gathered some evidence that deafness was hereditary and advised against the marriage of deaf persons because he believed a deaf race would result. However, at his time, there were certainly many other causes of deafness: Of 119 deaf children enrolled in one school, only 2 were children of deaf parents. Bell objected to "herding deaf students together" in residential schools and to deaf teachers. Although Bell was "fluent on his fingers," he urged that deaf students be taught through speech and speechreading.

In March of 1988, in events that will certainly be recorded in history as the deaf civil rights movement, students of Gallaudet University demanded and got the first deaf president of their university. As a result of their protests, a newly appointed hearing president was forced to step down.

Aids, Tickle Belts, and Bionic Ears

Many products of our electronic age are helping those with hearing disorders. Those who work with the hearing impaired have as their goal, they say, optimization of communication, including receiving information and developing language skills.

Hearing aids have gotten smaller and better. They are not a substitute for normal hearing, but they can amplify and filter sounds.

Through *vibrotactile* and *electrocutaneous* devices, the skin may become an auxiliary hearing aid, especially in distinguishing syllables and in distinguishing words that look alike on the lips. More than 100 such devices have been used, primarily to train deaf children in communications skills and to give them feedback about their own speech. The devices, of varying complexity, may be applied to any part of the body, taking into consideration the differential distribution of nerve endings over the skin.

One electrocutaneous device, the "tickle belt," provides electrical stimulation across the belly. The tickle belt mimics an uncoiled cochlea, with low sound frequencies represented on one end, high frequencies on the other. Children trained to speak with the aid of a tickle belt, which essentially gives them a word pattern on their stomachs, have been shown to produce more distinct syllables and to make "crisper" sounds than without it.

Cochlear implants, another product of the electronic age, can help those with sensorineural deafness by electrically stimulating the nerve endings of the inner ear. Single-channel implants have been done on hundreds of people since the 1960s, but multiple-channel implants—bionic ears—that selectively stimulate different areas of the cochlea have received a flurry of recent attention. The multiple-channel implants are up to 22 channels and, like calculators, are getting smaller. One of the greatest reported values of the cochlear implant is psychological—"the impact of not being deaf."

The American Medical Association has estimated that there are from 60,000 to 200,000 Americans who have suffered a profound hearing loss after learning language and who might benefit from a cochlear implant, but scientists are cautious about touting the multiple-channel implants as a replacement for hearing. Patients may feel less isolated, but the implants don't allow communication without lip-reading, and they don't provide an appreciation of music. The implants can help deaf people to modulate their own voices—to avoid speaking too loudly, too softly, or monotonously—and can help them to hear sounds in the environment, such as cars honking, dogs barking, or the telephone ringing.

The theory behind the cochlear implants is that if the stimulation patterns are the same, the brain will interpret them in the same way. The trouble is, however, that the cochlea has 30,000 separate nerve endings and it is physically impossible to stimulate them individually. Because comparatively few people have received the multiple-channel implants, and because success—measured in the recognition of spondee words, those with two stressed syllables—

has been variable, it is difficult to make generalizations about the implants.

The scientists are also cautious about recommending implants for children, because fibrous tissue can grow around them. Children might have extracochlear implants, they say, to avoid damage if cochlear implants are to be done later when they are adults.

The Sense of Communication

In a small town with a large extended family, I was awash in communication. When my grandmother quilted we sang and told stories. When her sisters came to help her get the quilt off, I played under the roof of the quilt and matched the different voices with the skirts and shoes surrounding me: serious Aunt Nora, shrill Aunt Johanny, soft-spoken Aunt Becky, silly Aunt Anise and Aunt Polly. Later I carried on conversations with my dolls and paper dolls, practicing.

When I visit my sister, my seven-year-old nephew, Bryan, often plays quietly nearby, making various noises and talking for his superheroes, using different voices. If there's a lull in our conversation, he looks up and asks, "Why aren't you talking?"

The windows of mine eyes.

—William Shakespeare

The harvest of a quiet eye.

—William Wordsworth

. . . I went thinking, thinking, wagging that human tail my cane, how all that I saw came to me thus only because of a specified convexity in the cornea of my eye. My sense of proportion, to say nothing of esthetics, is really superbly egotistic.

—Donald Culross Peattie

Born to see, made to view.

—Johann Wolfgang von Goethe

You can observe a lot just by watching.

—Yogi Berra

You're a sight for sore eyes.

—My mother

T · H · R · E · E

▼

Vision: A Viewpoint

VISION (from Lat. *videre*, to see), or SIGHT, the function, in physiology, of the organ known as the eye.

The Eyes: Extensions of the Brain

James Thurber, in his story "University Days," describes how in botany class he could never see cells through the microscope. His professor vowed grimly that his student would see cells. They would try every adjustment of the microscope known to man. With one of the adjustments, writes Thurber, ". . . I saw, to my pleasure and amazement, a variegated constellation of flecks, specks, and dots. These I hastily drew." The professor looked at his drawing, then into the microscope. "That's your eye!" he shouted. "You've fixed the lens so that it reflects! You've drawn your eye!"

The eyes are the only place in the body where a doctor can actually look inside and see the blood vessels. As Thurber discovered, sometimes under special circumstances we can see the reflection of the blood vessel network of our own eyes.

The eyes develop as extensions of the brain and continue to develop after birth. Newborns see only light and shadow. Later, they are farsighted, holding their rattles at arm's length to examine them. At four months, they see stereoscopically. Peak vision comes at about age six to eight.

The adult human eye is a slightly flattened sphere one inch in diameter, about the size of a Ping-Pong ball. The eyes sit within cone-shaped bony eye sockets, the orbits, formed by the joining of six bones of the skull. The orbits are padded with fatty tissue. Six straplike muscles, originating in the back of the orbits, hold each eye in place and move it finely and precisely, up, down, and around, 100,000 times a day. The eye movements allow us to follow a moving object while maintaining the object's image at a stationary position on the retina.

About one-sixth of the eyeball is exposed to the external environment, and the exposed portion is protected and nourished in several ways. The eyebrows and underlying brow ridges (more pronounced in our distant ancestors, who led more brutal lives) shade the eyes, protect them from blows, and collect perspiration. The eyelids, fringed with debris catchers, the eyelashes, shade and protect the eyes and cover them during sleep. The eyelids shut reflexively in response to rapidly approaching objects and sudden bright lights. *Meibomian glands* inside the eyelid produce an oily secretion that keeps the eyelids from sticking together (my mother found the secretions zoological—"sleepy bugs"—and others find them botanical—"sleepy seeds"). The *glands of Zeis*, at the base of the eyelash hairs, also produce an oily secretion that keeps the eyelashes supple.

The thick-walled and hollow eyeball is deceptively complex. The retina, the thin light-receptor layer inside the eyeball, has the highest rate of metabolism of any tissue in the body. In the embryo, the retina grows out from the brain to join the other tissues that form the eye. It is like a miniature peripheral brain designed to process visual information. In the retinal cells, energy is used in the rapid destruction and resynthesis of visual pigments, transmission of nerve impulses, and transport of substances across the retinal cell membranes.

The eyeball consists of three layers: the sclera, the choroid, and the retina. The *sclera* is the tough white of the eye, except for where it is transparent over the iris and pupil, forming the dime-size *cornea*. The cornea is curved and is responsible for bending about 70 percent of the light rays entering the eye; the lens, behind the iris, does the fine tuning.

The *choroid* layer contains blood vessels for nurturing the eye. It also brings nourishment to the *ciliary muscle*, which is attached to the lens, and the *ciliary nerve*, which enervates the ciliary muscle. The ciliary muscle alters the focusing capabilities of the lens by pulling on it and changing its curvature. The choroid is separated from the cornea by the two sets of muscles that form the colored part of the eye, the iris.

Light passes through an opening, the pupil, formed by the circular arrangement of the iris muscles. Pupil size is reflexively adjusted to varying degrees of light by nerve signals to the iris muscles. In dim light, the iris radial muscles contract, the circular muscles

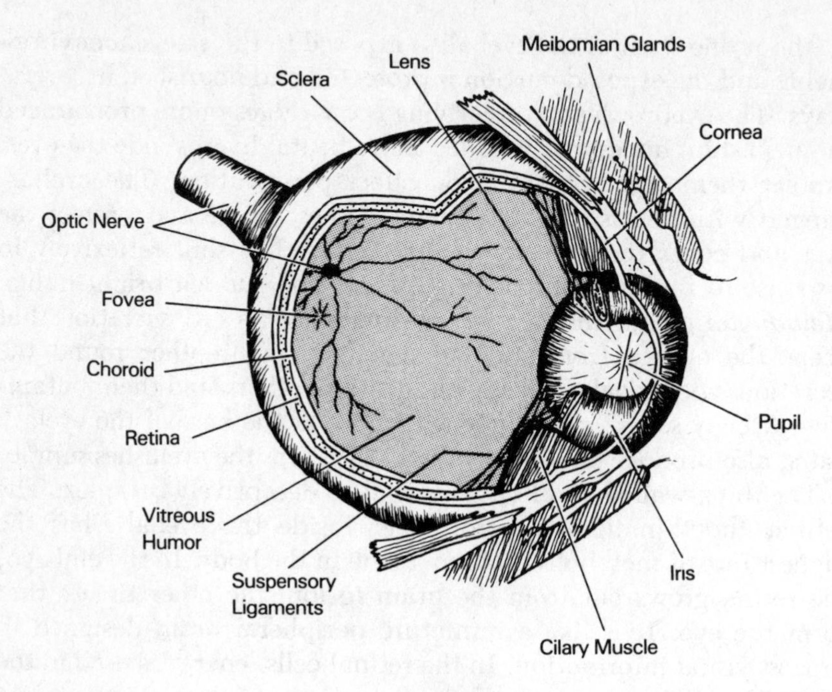

The human eye is an extension of the brain. You can see the reflection of your own blood vessels by looking at a blank wall in dim light and shining a flashlight or penlight into the pupil from the side. The vessel network will appear on the wall in front of you.

relax, and the pupil enlarges; in bright light, the opposite occurs. In darkness, the pupils may be as large as a pencil eraser; in bright light, they may shrink to the size of a pinhead.

Pupil size is also affected by emotional states, as we shall see; the windows to the soul is a biological as well as a poetic concept.

The retina has three layers: pigment cells, light-sensitive cells, and switching or processing cells. The pigment cells serve the same function as black paint on the inside of a camera body: They absorb light that would otherwise bounce around in the eyeball, causing blurring. (Not all animals have the dark pigment. Animals that are active at night have a highly reflective layer known as the *tapetum*, sacrificing clarity of image for the chance to make use of a limited amount of light. A cat's or a deer's eyes glow when it looks at a car headlight beam because of light reflected from its tapeta.)

The receptor layer of the retina is made of cells called *rods* and *cones* because of their shape. Microscopically, the rods look like cattails on a stalk; the cones, responsible for color vision, are thicker at their base, like carrots. Rods, which distinguish only light, are much more numerous than cones in the human retina, about 130 million to 7 million. The cones are concentrated at a yellowish spot, the site of focus of the central ray of light entering the pupil. This spot of concentrated cones, the *macula lutea*, contains a pit, the *fovea*, made up entirely of cones; it is the area of greatest visual acuity and discrimination. Rod and cone cells extend outward from the retina and communicate with the *ganglionic neurons*, whose processes converge to form the cable of the optic nerve.

The optic nerves exit each eye socket then converge at the *optic chiasma*, where some partial switching takes place. Neurons from the nasal side of the right eye travel to the left part of the brain, and neurons from the nasal side of the left eye travel to the right part of the brain. This arrangement results in the left side of the brain getting information from the two left retinas, and the right side of the brain getting information from the two right retinas, and is presumably useful for the brain in combining information received from two different points in space. Visual information carried by the optic nerves is interpreted at the back of the brain, in the *occipital lobe*. A blow to the back of the head can make a person "see stars," and damage to the occipital lobe can alter vision as surely as damage to the eyes themselves.

The point at which the optic nerve leaves the eye is the *optic disk*. This is the *blind spot*, so named because it has no light receptors. We are normally not aware of our blind spots—we fill in the missing information with our brains, but we can easily demonstrate its existence.

For our prehistoric ancestors, the eye's main job was to see things at a distance—danger to be avoided, game to be killed. Only comparatively recently has the human eye been called on for continuous close-up work. It takes more muscle work to see near objects than to see clearly into the distance.

The lens of the eye, which is made of layers of protein fibers, accommodates for near vision by muscle action. The tiny muscles surrounding the lens constrict, like pulling drawstrings on a bag, and the suspensory ligaments are loosened, causing the lens to shorten and thicken. For far vision not as much bending of the incoming light rays is required. The ciliary body muscles are relaxed, and the lens is flatter.

Those of us who do mostly close work, such as writers, can rest our eyes periodically and let the ciliary muscles relax by looking into the distance. Changing the pattern of eye movements from looking from word to word on a page all day long to large movements of the eyes also improves circulation. After large eye movements, it becomes easier to look at small detail again. This is a good argument for large offices (newsroom style) or offices with windows. I'm not sure we're conscious that we're giving our eyes a rest, but I never hear office workers or students say they like windowless rooms. I've heard complaints about small rooms with no windows, and I notice that the people in my office building have all placed their desks near windows so they can periodically gaze into the distance. The most popular study spots in my university library are near windows.

The lens continues to grow during a person's lifetime, losing its elasticity and becoming more difficult for the tugging muscles to distort. As the ability to accommodate for near vision is lost—a condition known as *presbyopia*—people begin to hold newspapers at arm's length to read them and realize they could benefit from reading glasses.

The lens separates the hollow interior of the eye into two liquid-filled cavities. In front of the lens is a watery *aqueous humor*, a filtrate of blood plasma that is constantly renewed. Back of the lens

Find your own blind spot or optic disk, the area on the retina where the optic nerve leaves the eye. Cover your left eye and stare directly at the cat at the upper left. Move the book toward and away over a range of about 12 to 40 centimeters. At one point, the dog will disappear. Do the same thing with the cat at the lower left. The dog, but not the stripes, will disappear.

the eye is filled with the *vitreous humor*, a glassy fluid the consistency of egg white. The vitreous humor is made early on in embryonic life and is never replaced. It contains microscopic particles, pieces of shed cells, that float freely. These "floaters," or "flying gnats," are visible when we look up or to the side. I especially notice my floaters following the lines of type when I read in bed, lying on my side.

The Exciting Visible Spectrum of Light

Every object we see, we see as a result of light energy, either light emitted by the object or light reflected from it. *Visible light* is a form of energy called electromagnetic radiation. Only a small fraction of the entire spectrum of electromagnetic energy consists of visible light. The visible light spectrum is bracketed with rays of shorter wavelengths and higher energy, such as ultraviolet rays, X rays, and gamma rays, and with rays of longer wavelengths and less energy, such as infrared, radio, and radar waves.

The Greek philosopher Pythagoras (about 600 B.C.) thought light was a steady stream of particles. Aristotle (384–322 B.C.) thought light traveled in waves. Both were correct. Curiously (for our commonsense brains), light has characteristics of both particles (one particle is called a photon) and waves, depending upon how it is measured.

Light travels in straight lines, except for diffraction (the spreading around an obstacle), and it behaves in predictable ways. In the first century, Hero of Alexandria discovered that light striking a mirror, a flat polished surface, is always reflected at the same angle; that is, the angle of incidence is always equal to the angle of reflection (the same law of mechanics operates on a pool table, with reflecting balls). This property of light allows us to manipulate it and to observe an image of an object at a location removed from the object. Most often, probably, we use it to look at ourselves in the bathroom mirror every morning. (And what we see in the mirror is opposite of what other people see when they look directly at our faces.)

Many more properties of light were illuminated 2,000 years later, in the seventeenth century. People had long observed the apparent bending of an object placed in water at an angle. The Dutchmen Willebrord Snell and Christian Huygens studied the bending of light and worked out the refractive indexes of different materials—the ways that different materials can be expected to bend light.

Sir Isaac Newton (1642–1727) discovered that light directed through a prism in a dark room produced a series of colors, beginning with red at one end and proceeding through orange, yellow, green, blue, indigo, and violet. The colored beams, he found, could be redirected through another prism and recombined into white light. White light is made up of the rainbow of colors. Newton also isolated the colors and found that nothing could be done to change any of them. Each has its own wavelength.

It is no accident that living things "see" in the spectrum of visible light with wavelengths between about 380 and 750 nanometers (a nanometer is one-billionth of a meter and is a measurement of the

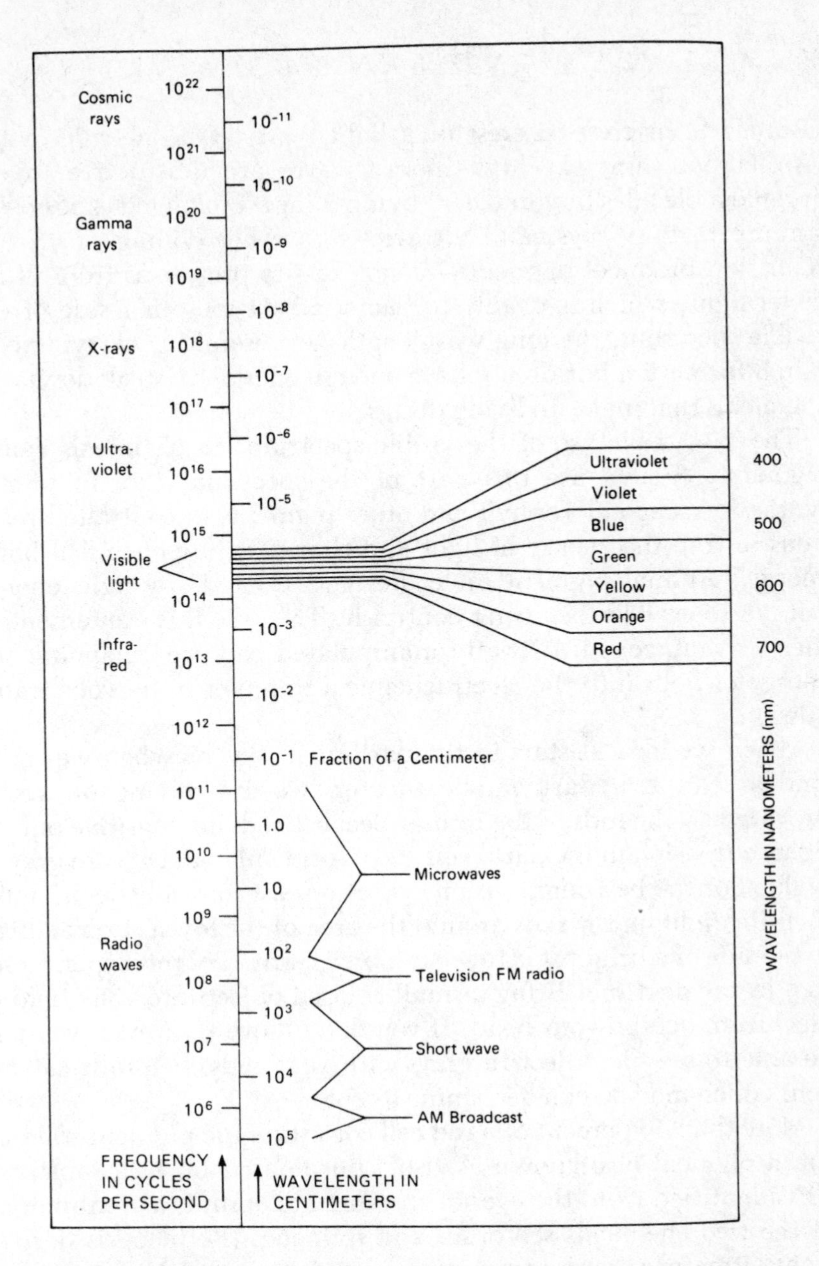

Visible light is only a portion of the energy of the electromagnetic spectrum. Light coming from the sun contains about equal amounts of the different wavelengths, which we perceive as white. We perceive the different wavelengths as different colors. Most plants absorb wavelengths in the far left and far right parts of the spectrum and reflect the wavelengths of 550 to 600 nm, so we see plants as green, from the light they reflect. Black is the absence of light.

distance from crest to crest of a light wave). On one side of the visible spectrum, the high-energy waves are destructive to the organic molecules that make up living things. Prolonged exposure to gamma rays, X rays, and ultraviolet rays has damaging effects. (Our skin produces pigments—a suntan—to protect us from ultraviolet light, which is deadly to bacteria.) At the other side of the visible spectrum, the long wavelengths are useful for carrying certain information but do not have enough energy to break down the chemicals that make up living things.

The eyes make use of the visible spectrum via chemicals called pigments. (Plants use this part of the spectrum, too, in photosynthesis, using chlorophyll and other pigments in controlled reactions to trap the energy of light and change it into chemical bond energy.) Animal pigment molecules get "excited" by light energy but not so excited they can't control it. They use this excitement to their advantage, in a strictly manipulated capture, changing the energy of light into the electrochemical impulses of nervous transmission.

When we look at stars in the dark night sky, or when we try to find a seat in a dark movie theater, we are relying on vision provided by the rods of the retina. Because rods are sensitive only to light, our vision in the dark consists of form only; all cats are gray at night. For the best night vision, we also must look a little askance, focusing light on the rods around the area of the fovea. Look a little to the side of a bright star to see it more clearly. Try this experiment, too, in the daytime: Bring a small colored object into your field of view from behind your head. If you don't move your eyes, you will be able to see the object in gray, with your rods, before it catches your cones and you can determine its color.

More than 40 percent of a rod cell consists of the pigment *rhodopsin*, a pigment also known as visual purple (similar to a substance first identified from the eyes of frogs). Rhodopsin is a combination of the two chemicals scotopsin and retinene. (Retinene is derived from vitamin A; carrots *are* good for night vision.) Night vision is called *scotopic vision*.

In bright light, rhodopsin is quickly broken into its component chemicals, stimulating nervous impulses in the rods. Much rhodopsin remains bleached out in the daytime. When I enter a dark movie theater I must wait until my eyes resynthesize enough rhodopsin to react to dim light. Then I can see form well enough to avoid sitting on someone else's lap. The dark-adapted eye is most sensitive to photons of light at the 505-nanometer wavelength. If it is a daytime movie, when a person goes back into bright light, rhodopsin is broken down rapidly, sending many signals to the brain at once, making it difficult to see. After most of the rhodopsin is bleached out, fewer impulses are transmitted to the brain; the eyes have become light-adapted.

People with night blindness do not have normal night vision following the adjustment period. Night blindness may be a genetic condition or may be caused by vitamin A deficiency and lack of retinene, so that no pigments are present to react to weak light rays.

Humans live in a world of light and dark—stars and movie theaters—but we also live in a wonderful world of color (that is, the 95 percent of us who are not color-blind; we'll get to that). We have *photopic vision* as well as scotopic vision. Color, however, is purely psychological, an interpretation of our brains. Light rays are not colored. Our physiology has made them correspond to different colors.

By selectively blocking bands of the light spectrum, and recombining them, Newton observed that if as few as three basic colors were present, white light resulted. These are the primary colors: red, green, and blue. Adding them back together selectively produces different colors in a predictable way. Newton projected and superimposed colors on a screen to work out combinations.

The color television screen is doing something similar, adding colors of different wavelengths by placing them side by side in close proximity. If the television patches are small enough, the human eye is unable to separate them and will respond as if the lights are superimposed. Look closely at the color television picture; it is made up of minute fields of red, green, and blue. When you step back, the individual colors merge into additive pictures of the three colors.

In 1807, based on knowledge about the behavior of visible light, rather than detailed knowledge about the physiology of the human eye, the English physician and physicist Thomas Young (1773–1829) proposed that the human eye must have three fiber types, one sensitive to red, one to green, and one to violet. (Young was interested in figuring out other puzzles too; he helped translate the Rosetta Stone.) Young's trichromatic theory was formalized in the

Sir Isaac Newton made monumental contributions to physical optics and mathematics and is regarded as a genius who discovered many principles of the workings of the world. Newton used prisms to separate light into its component wavelengths. A rainbow is produced by the light-separating effect of each individual raindrop, which acts like a prism. The curvature of each drop refracts the various wavelengths and spreads them into the color spectrum.

mid-1800s by Hermann von Helmholtz (1821–1894) in *Physiological Optics*. Helmholtz proposed hypothetical excitation curves for each of the three types of fibers in the eye and used these curves to predict the ability of humans to make color discriminations on the basis of wavelength. (In 1851, Helmholtz invented the ophthalmoscope, an instrument that proved to be a most valuable contribution to understanding the physiology of the eye.)

The actual measurements of the excitation curves of the cone pigments were done later, in the 1960s, by scientists using a microspectrophotometer, an instrument that can focus a fine beam of light on an individual human cone cell. Three types of cones were found, those with absorbance maxima at 450 nanometers, those at 525 nanometers, and those at 555 nanometers, corresponding with blue, green, and red wavelengths. Young was right. Each individual cone contains only one of the three cone pigments, each excited at different wavelengths. The combination of excitation is interpreted by the brain as color. Humans have what is known as a trichromatic color vision system; the best there is. (But before we get smug, I should mention we share this trait with many other animals, including a carp.)

I See, I See

Euglena is a one-celled organism that lives in pond water. It comes in many species. Unlike the ponderous amoeba that is easy for students to follow under the microscope lens, *Euglena* whips itself around quickly with a long, hairlike flagellum. It has a red eyespot and will swim toward ordinary daylight but away from the direct rays of the sun. (Because it will absorb nutrients from the water, and because it also has the pigment chlorophyll and can make its own food, it is presented to biology students so this unanswerable human question can be asked and pondered: Is *Euglena* an animal or a plant? Of course, it doesn't matter to the *Euglena*. It is what it is.)

Almost every living thing is sensitive to light. The first simple eyes responded only to the changing intensity of light. Perception of form and color awaited multicellularity, more complicated eyes capable of forming images, and brains able to interpret signals from optical images on the retinas.

An earthworm has light-sensitive cells scattered over its skin, but in most animals the light-sensitive cells are arranged in groups, often lining a depression or a pit, which is the beginning of a true image-forming eye. A simple pit eye is still found in limpets. The pit permits the animal to detect the direction of incoming light. For the same reason, Greek astronomers dug deep holes in the ground from which they could observe stars in the daytime, eliminating side-scattered light not coming from stars.

The development of a transparent membrane over the pit eye protected it from debris. Chance mutations that made the membrane thicker in the center became a crude lens. (The nautilus took a different route. It has an eye like a pinhole camera, with a tiny pupil that allows the eye to be washed by the sea in which it lives. Eyes with lenses must specially manufacture the fluids to replace the sea. The nautilus eye is in focus at all distances yet admits very little light.)

Nature has many ways of looking at a blackbird. Scientists who study eye designs have concluded that photoreceptors of various degrees of differentiation have evolved independently in at least 40, and maybe 65, separate groups of animals, testimony to the importance of light perception. Complicated eyes may go with fairly simple brains. The compound eyes of the insects and their relatives consist not of a single lens but of many tiny individual eyes, known as *ommatidia*, each connected to its own nerve ending. The eye idea was well developed with the trilobites, who crawled the seas 600 million years ago but exist today only as fossils. The eyes of some trilobites had more than a thousand facets. Today, some insect eyes have 30,000 facets. The insect eye as a whole does not move, and the lenses cannot be focused. Each facet catches a small piece of the surrounding scene and the individual pieces are combined into a single picture. The more facets its eye has, the more detailed a picture an insect can perceive. The compound eye works especially well at short distances.

The eyes in the insect and spider world are present in an incredible arrangement of shapes and sizes and sensitivities. The dragonfly has wraparound eyes. Bees see colors, especially yellow, and they also see ultraviolet light. Many bee-pollinated flowers appear much differently in ultraviolet light, with "bee guides," landing pad patterns visible to bees but not to us.

Swimmers know that vision in water is different from vision in air; in water, light is diffused and quickly fades into a twilight zone. Longer wavelengths of light are filtered out. Whirligig beetles, those little black beads (one kind of "Jesus bug") that gyrate in groups on the top of pond water or still parts of streams, have two sets of eyes, an arrangement I have often heard teachers and parents of young children wish for. One pair, adapted for vision in the air, keeps watch for airborne predators such as birds. The other pair, adapted for vision in the water, keeps an eye out for fish. A small fish of Central and South American rivers, the anableps, has a similar arrangement, but the differentiation is within each single eye, and the fish swims with the upper halves of its eyes out of the water.

Most spiders have eight eyes, apparently the original number for them, but various lineages have lost some, so there are six-eyed, four-eyed, two-eyed, and even a one-eyed cyclopean spider. The salticids, the jumping spiders, are big-eyed daytime hunters that can see green, blue, and ultraviolet. Life in the sun, writes spider expert Willis Gertsch about the jumpers, has produced a variety of brilliance of coloration unmatched by other spiders; a display of

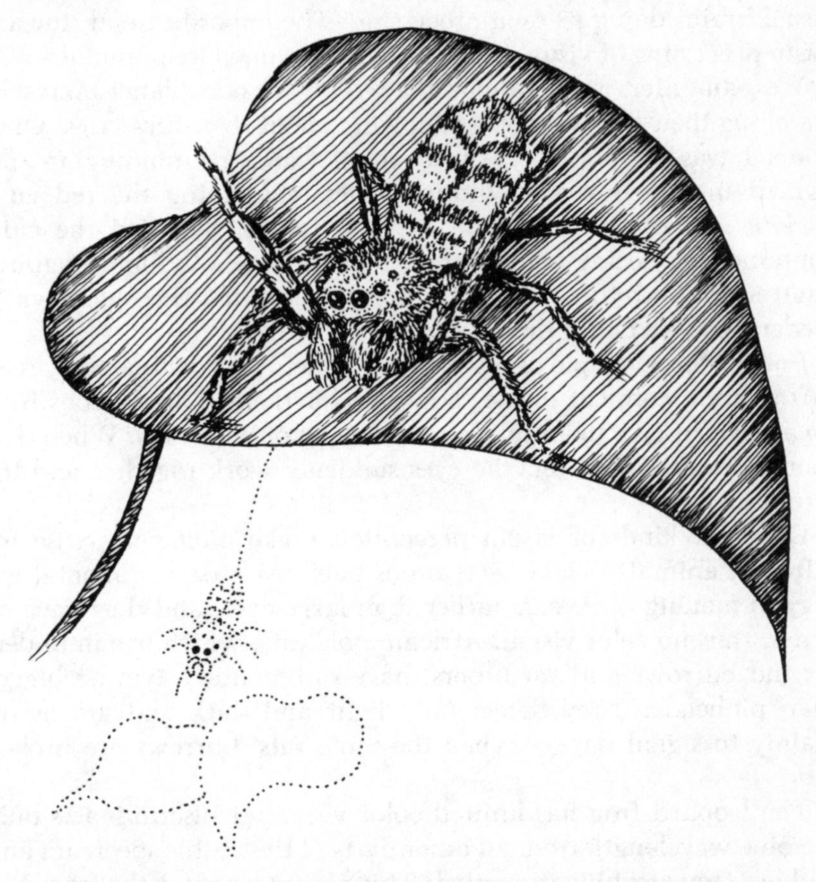

The eight-eyed jumping spiders, active daytime hunters, are some of the most easily recognized. They can receive a sharp image at a distance of 10 to 12 inches with their large eyes, and perceive moving objects at much greater distances with their smaller eyes. They leap and dance gracefully, saving themselves from falls by dragline threads. They will leap away from a building to catch insects in flight.

this ornamentation is part of the spiders' courtship ritual. Jumping spiders are the spiders with personality. A jumping spider looks back at you. It will sit upon your finger and watch your every move.

Our own eyes are a good example of the vertebrate eye design. Vertebrate eyes are long-range instruments, but within the vertebrate group are many specializations. The dinosaurs, for instance, are thought to have been highly visual animals, but they had small brains. In dinosaurs the retina probably functioned as a complete visual brain, doing its own processing. The opposite trend, toward brain processing of visual information, developed in mammals.

We can infer much about the vision of our planet-mates by watching their behavior or by knowing about their life-styles. Once when I was out for an early morning walk, a hummingbird delighted me by hovering within inches, examining the red cuffs sticking out of the sleeves of my tan jacket. Satisfied the cuffs contained no nectar, the hummer buzzed away. As hummingbirds chatter around the trumpet vine flowers and the sugar-water feeders, can we doubt their penchant for red?

I once kept a couple of chameleons. Their uncoordinated eyes on turrets were comical to me, adaptive for them. Chameleons have scanners that give them split-screen images of the world. When they spot an insect, however, the eyes suddenly work together and the screen merges.

Different kinds of visual perception make biological sense for different animals. The insectivorous bats invest developmental energy in making large ears rather than large eyes, and they have no cones, thus no color vision. African mole rats, which live in underground burrows and eat tubers, have rudimentary eyes no bigger than pinheads. They detect only light and dark and are useful mainly to signal danger when the mole rats' burrows are broken into.

The leopard frog has limited color vision. It discriminates only the blue wavelength from all other parts of the visible spectrum and will leap toward blue if startled. This makes sense for the frog: If a frog is sitting on a lily pad, catching insects, blue is water, which equals safety. Frogs also have "bug detectors" built into their visual systems. Frogs react to small movement by thrusting out their sticky tongues. We demonstrated this in biology laboratory by dangling carrot pieces on string; but small movement in a normal frog's life usually means a flying insect. Bug detectors don't work on dead insects. A frog will starve to death in a room full of dead flies.

Birds have extremely big eyes (an ostrich's eyes are the biggest, a monstrous two inches in diameter), weighing more than their brains; and daytime-feeding birds have retinas rich in cones. Birds use bright colors to communicate with each other; note the male peacock's tail display or red-winged blackbirds flexing their epaulets in the spring. Owl eyes, on the other hand, are rich in rods, for night vision. They are almost color-blind. No fancy tails. No epaulets.

Different birds, and other animals, too, have different fields of view and different combinations of monocular (one-eyed) and binocular (two-eyed) vision, depending on the positioning of their eyes. In general, hunters find useful the depth perception provided by binocular vision; the hunted find it more adaptive to have a wider field of view so they can watch for hunters. (Close one eye and look at your hand. Close the other eye and look at it. When viewed with one eye, it appears flat. With both eyes, it becomes a three-dimensional object in space.)

Pigeons, with lateral eyes, have a total view of about 340 degrees; they can see everywhere but directly in back of the head. Their binocular vision, however, is only about 24 degrees of their field of view. Hawks and owls with eyes set forward have a smaller range of side vision but a wider range of binocular vision: from 35 to 70 degrees in front of them.

Many arrangements are possible. Woodcocks, with their eyes almost on the top of their heads, can see backward and upward, and forward and upward, with binocular vision and laterally almost 180 degrees with each eye. Bitterns have their eyes placed low on their heads and can see beneath their bills. They live among reeds and can escape notice by freezing with their bills pointed skyward. Holding that pose they can still look forward binocularly.

The hawks and eagles have eyes about the size of a grown man's, with two foveas on their retinas. One, the search fovea, functions in lateral vision; the other, the pursuit fovea, functions in binocular vision. The human fovea has about 200,000 visual cells for each square millimeter; the fovea of the North American red-tailed hawk contains one million visual cells per square millimeter, five times as many as humans.

Birds have faster vision than humans—that is, they pick up details more quickly. With a single glance, a bird takes in a picture that a human accumulates by scanning the field of view piece by piece. Birds also have greater sensitivity in picking up movement. Unlike human eyes, bird eyes are fixed in their sockets. They turn them by turning the head and neck. The red, red robin goes bob, bob, bobbin' along because it lacks extrinsic eye muscles but makes up for it in neck muscles.

An owl has a stationary field of vision of about 80 degrees but can rotate its head 270 degrees, about three-fourths of a full circle. Quickly turning its head back and forth, an owl can see all around it.

The visual acuity of primates probably developed in tree life, allowing fine discrimination of ripe fruits among the green leaves and of other primates. Nocturnal primates usually don't see color, and the New World monkeys have trouble distinguishing reds and greens; but the primate group in general has good color vision. The living primates present an array of colors and conspicuous patterns useful in social communication. The color vision adaptive for tree life was undoubtedly helpful—preadaptive—for our ancestors on the savanna, helping them to see small animals that might have been more difficult for color-blind predators to detect.

In the primate evolutionary history, eyes have moved forward on the face so that we have good stereoscopic vision and depth perception. Our human field of view is about 180 degrees, half a circle, and of that the overlap is about 150 degrees.

Color Defects

Many humans have defects in their color vision, so that their visual worlds are different from the visual worlds of the rest of us. Some have trichromatic vision but have a cone type with abnormal pigments. An eye examination reveals that anomalous trichromats have a weakened ability to make certain color discriminations, especially in the green and red areas of the spectrum.

Many other types of color blindness have been identified, but color blindness is usually a sex-linked hereditary defect, associated with a recessive gene on the X sex chromosome. Because human females have two X sex chromosomes, while males have only one, the trait is more common in males. (A female must have two recessive genes to express the color-blind trait; males need only one. Color-blind males get the trait from their mothers, who are the donors of the X sex chromosomes. Fathers donate the Y sex chromosome to their sons.) Eight to 10 percent of human males have some color vision defect, compared with only 1 percent of the female population.

The most extensively studied color-blind people are the *dichromats*, who are completely missing one kind of cone pigment. They may be *protanopes* (lacking the long wavelength-sensitive pigment), *deuteranopes* (lacking the medium wavelength-sensitive pigment), or *tritanopes* (lacking the short wavelength-sensitive pigments). Depending on the deficiency, the area of color confusion will differ. Human achromatic vision—total color blindness—is extremely rare.

Color blindness was described by the English chemist and physicist John Dalton (1766–1844), who realized when he was 26 and

John Dalton, best known for his contributions to atomic theory and for creating a table of atomic weights, put his scientific bent to work as a young man to describe his own color blindness, now known as protanopia. Dalton is said to have scandalized his Quaker congregation by wearing crimson socks.

had been studying botany that he and his brother had some visual peculiarities; prior to that time, he had just thought there was some "perplexity" in the naming of colors. Geranium flowers, for instance, appeared to him sky blue in the daylight, while others called them pink. By candlelight he observed that they seemed yellow with a little red. No one else, except his brother, Jonathan, saw them that way.

In 1794 Dalton was elected a member of the Manchester Literary and Philosophical Society, and a few weeks after his election contributed his first paper on "Extraordinary facts relating to the vision of colours," in which he gave the earliest account of what came to be known as Daltonism or color blindness. He and his brother had protanopia, the most common kind.

Dalton began examining the color vision of others with tests using colored ribbons and questions about commonplace things such as grass, sky, and blood. He found that most people distinguished six colors of the solar spectrum: red, orange, yellow, green, blue, and purple; but he also found about 20 males who had peculiarities of color vision. Dalton found that what he called blue and purple coincided with reports of those with normal color vision, but what he called yellow included the red, orange, yellow, and green of others with normal vision.

Dalton had low sensitivity in the red part of the spectrum, seeing ruddy complexions as dusky blue. Pink for him was composed of about nine parts of light blue and one of red. His observation of the geranium in candlelight is explained by the fact that candlelight gives off more light energy to be reflected in the red rather than blue end of the spectrum, so he saw the flower as red, which for him was a murky absence of sensation.

Dalton noted that the color vision peculiarity was more common in males and that it tended to run in families. He speculated that his eyes contained a blue fluid that absorbed green and red light. Despite Thomas Young's writing that Dalton's color defect might be due to the absence of a set of vibrating fibers in the eye, Dalton insisted that upon his death his eyes be removed and their fluids examined for blueness. His wish was carried out 50 years after the speculation, and Dalton's eye fluids were found to be normal, colorless.

By the middle of the nineteenth century when Dalton died, color blindness as an individual curiosity was generally known to the public at large, and with the installation of red and green color signals for railways, tests for color blindness were made mandatory for railroad engineers.

The Eye as Camera, and Optograms

The working of the eye and its interaction with light and the objects we see occupied the thoughts of many natural philosophers. It was an awesome problem. About 500 B.C., the Greeks developed a theory that vision resulted from lightlike rays that streamed from the eyes, touched an object, and made it visible. (Superman, with his X-ray vision, still carries on this kind of misconception.) Aristotle saw the flaw in the idea. He asked, if eyes are the source of light, why are things invisible in darkness?

But humans found it difficult to think in simple terms about light and eyes. Misconceptions held back the understanding of the eye for many centuries. The gross optics of image formation were finally expressed by Johannes Kepler in 1611 and again by René Descartes in 1664.

The eye is like a camera in many ways. Both devices project an inverted image of the surroundings upon a light-sensitive surface: in the eye, the retina, in the camera, the film. In both, light admitted is regulated by an iris. The eye is focused by changing the thickness of the lens, the camera by moving the lens toward or away from the film. The change from cone to rod vision is like changing from fast to slow film, from color to black-and-white.

The use of a lens to project an image creates problems for both eye and camera: spherical and chromatic aberration. Spherical aberration brings the marginal rays of light to a shorter focus than the central rays, and chromatic aberration results from the fact that short wavelength rays are bent more strongly than long wavelength rays. Cameras correct for both problems by modification of the lens. The cornea of the eye corrects for spherical aberration by a flatter curvature at its margin. The eye addresses color aberration in several ways: The yellow lens cuts off ultraviolet rays (people who

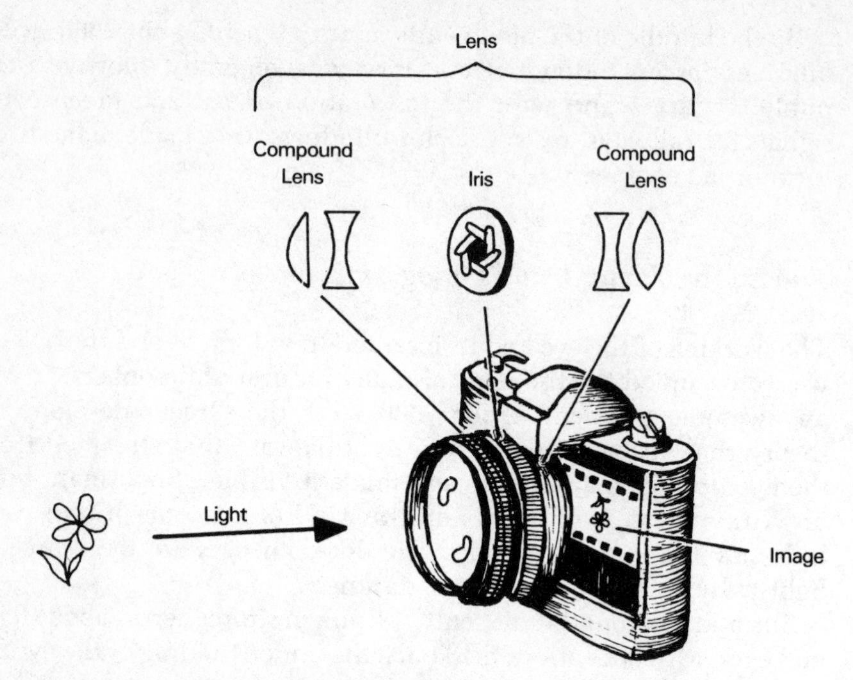

Both eye and camera use a lens to focus a reversed and inverted image on a light-sensitive surface. Both have an iris to adjust to various intensities of light. The single lens of the eye, however, cannot bring light of all colors to a focus at the same point. The compound lens of the camera is better corrected for color because it is composed of two kinds of glass. The upside-down and backward image posed problems of explanation for natural philosophers, but the brain has no problem righting and interpreting it.

have their lenses removed because of cataracts can see in the ultraviolet), and the yellow pigment in the macula lutea, the area of most acute vision, further cuts off shorter wavelengths.

In the late 1800s one scientist, Willy Kuhne, from his studies of the action of the chemical rhodopsin, saw the possibility of taking pictures with living eyes, "optograms." Among other things, he exposed the eyes of a rabbit to a barred window, then killed the rabbit, removed its eyes and fixed its retina, upon which was seen the light and dark pattern of the bars. In 1880 he arranged to obtain the eyes of a man who was beheaded, and from them he printed an optogram that does show an image, but one that is impossible to interpret.

With both eye and camera, the light rays are bent toward each other and intersect, so that the image formed on the film, and on the retina, is upside down and backward. I once saw a film in psychology class in which a researcher wore a pair of goggles with lenses that reversed and inverted the image so that what came to the retina was also upside down and backward. For a while this is how he saw the world. Later, surprisingly, the brain flipped the image so that he saw the world right side up. When he took off the goggles, his world was upside down again. It took a while to get back to normal, and even then, sometimes his brain made some spontaneous flip-flops. This kind of experiment has been done several times. The narrator of the film said, essentially, this is a very interesting experiment, but don't try it yourself. (Read *Guinea Pig Doctors* by Jon Franklin for more amazing stories of self-experimentation.)

We See What We See

The analogy of eye as camera stops in the photographic laboratory and in the filtering of optical messages by the retina and interpretation by the brain. Babies learn to associate certain spatial relations in the outside world with certain patterns of nervous activity in the eyes. The spatial arrangements of the nervous activity itself—the upside down and backwardness—are irrelevant. The brain translates the information into something meaningful. How it does it is something else altogether.

Researchers have spent a great deal of time trying to decipher just how the brain makes sense of the signals it receives from the retina. We do know that it makes assumptions about what should be right and that it can be tricked. Magicians use smoke and mirrors, but there are lots of other classical optical illusions. Some are not what they seem, others are impossible figures, others are ambiguous figures.

Seeing is actually being deceived. Consider reading. As you read this, your nose intrudes into the field of view, but you ignore it. Despite blinking, we don't perceive the world as flickering light. We have learned and internalized the physics of the everyday world and have a representation of it in our minds. We know that space is

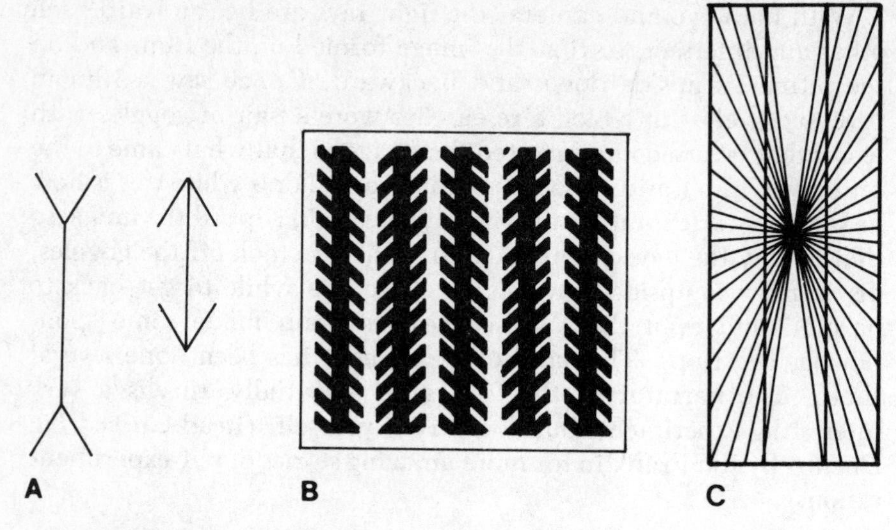

Optical illusions, figures that lead to erroneous perception, help us to hypothesize about how the brain accepts and processes visual information. The vertical line on the left in A *appears much longer than the vertical line on the right, because the diagonal lines cause the viewer to make an incorrect comparison. In* B *and* C*, diagonal lines also distort perception of the vertical lines.*

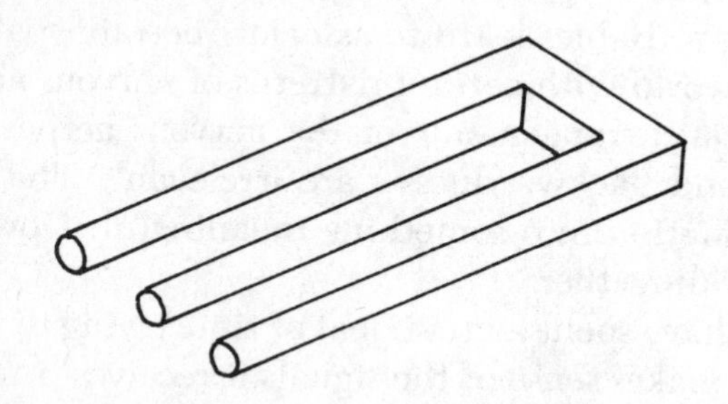

When we first look at this drawing of an object, it appears reasonable, but upon closer inspection we see that something is wrong. The artist M. C. Escher is famous for creating complicated impossible figures.

three-dimensional, that things have edges, that up and down exist, that falling objects are predictable. Much of what we see is what we learn to expect to see. Magicians say they trick others by getting them to come to the wrong conclusions for the right reasons.

Vision is not a single process. It is keeping scientists busy in many laboratories, exploring many different aspects of how we see what

This ambiguous figure, redrawn from an illustration by W. E. Hill, demonstrates the importance of interpretation in the process of perception. A young girl or an old woman can be seen, depending upon the viewer's perspective. The young girl's chin is the old woman's nose.

we see. A unifying theory of vision is not in the offing, but understanding aspects of visual information processing holds promise for a variety of applications, especially in computer science and artificial intelligence.

Lenses by Nature and Humans

When Chris and I, both sans our eyeglasses, were looking at a magazine pinup (John-John Kennedy) in the grocery store checkout line, she held the magazine at arm's length. I held my nose close to the page, while trying not to obscure her view. We have opposite problems: She's farsighted (hyperopic); I'm nearsighted (myopic). She puts her glasses on when I take mine off, and vice versa. In farsighted people, the converging light rays, the focus point, extends beyond the retina; in the nearsighted, they converge in front of the retina. Chris wears her glasses to see close objects. I wear mine for distant vision. My eyeballs are a little too long; hers are a little too short.

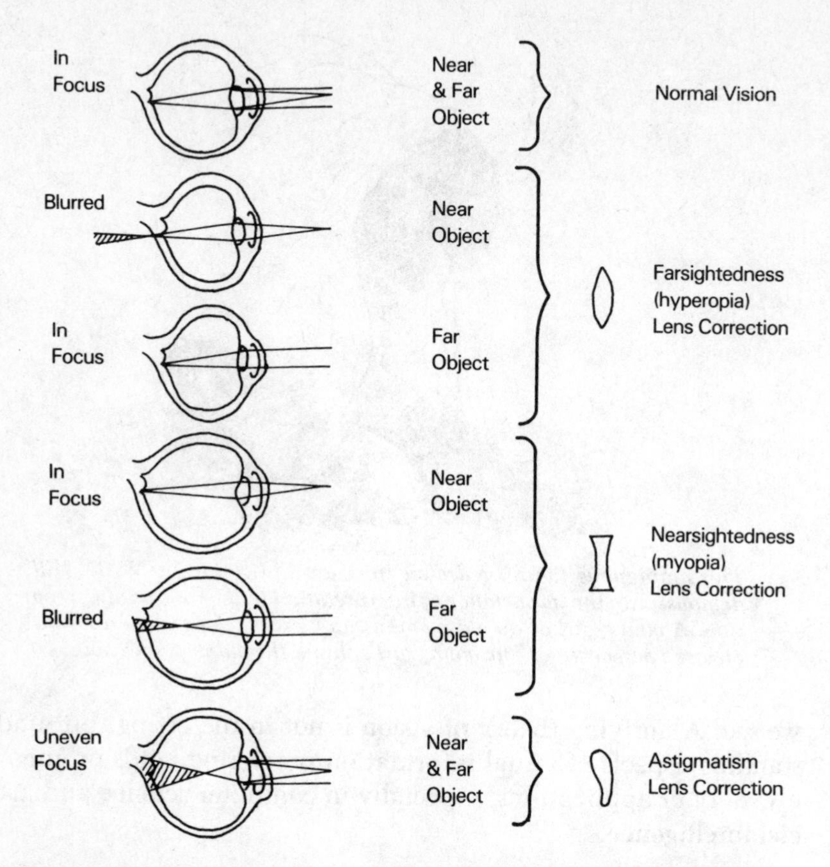

In normal vision, light rays are focused to converge at the retina. In farsightedness, or hyperopia, the incoming light rays come to a focus behind the retina. The problem is corrected by a biconvex lens. In nearsightedness, or myopia, the incoming light rays focus in front of the retina, and a biconcave lens is used for correction. In astigmatism, the cornea or lens is curved unevenly, and the problem is corrected with a lens with the same degree of astigmatism in the opposite position.

Neither of us has astigmatism, however. In astigmatism, the cornea or lens is uneven or bumpy, so light rays are focused unevenly on the retina, resulting in fuzzy or distorted vision. Astigmatism is corrected by an unevenly ground lens.

Although only 2 to 3 percent of our population is born with defective vision, half of us eventually wear corrective lenses, and many more live with undetected vision problems.

Early hunters must have depended heavily on keen distant vision. Those in the population who were extremely myopic may have found their niche as tool crafters. Many probably did not live long enough to develop presbyopia. The human eye was meant for distant vision and for gradual movements, yet the demands of culture have put different stresses on the visual system—direct stress on the lens system as well as upon the visual cortex of the brain where images are interpreted.

In fact, reading seems to contribute to myopia. In 1813, James Ware, a British officer in the Queen's Guard, noticed that his well-educated fellow officers were frequently nearsighted, while the men under his command, virtually all illiterate, almost never were. A 1960s study of Eskimo in Barrow, Alaska, found that only 2 of 130 Eskimo parents were nearsighted, while 60 percent of their children

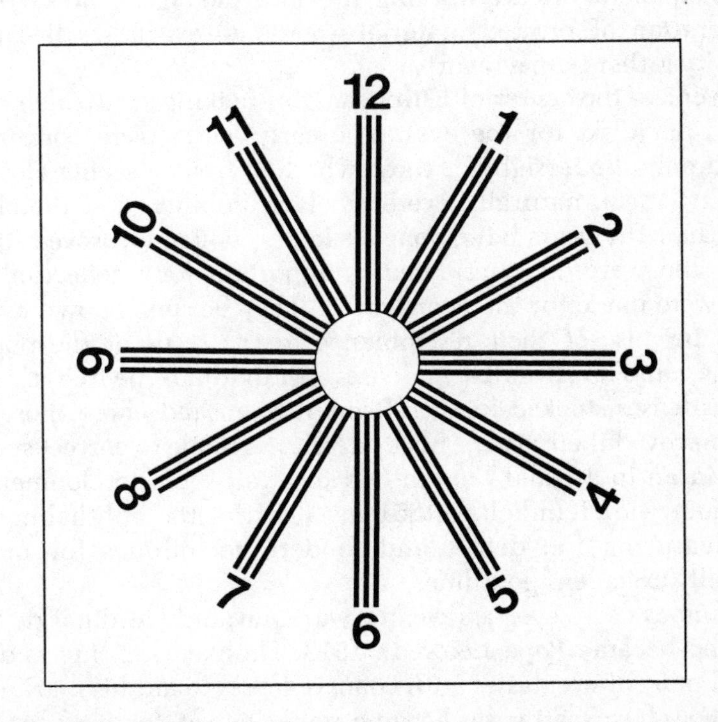

Even minor amounts of astigmatism can be detected with the standard testing chart. To a person with an asymmetrical cornea, lines in one plane appear sharp and clear while those at right angles are fuzzy.

were. The principal difference between the Eskimo groups was compulsory schooling, that is, reading.

Recent experiments with chicks have revealed how reading and myopia are related. The chicks weren't required to read, but rather wore contact lenses that were opaque on the outer part, obscuring their outer field of vision. Understimulation of the peripheral part of the retina resulted in lesser activity of those cells, which, combined with greater activity of the cells in the center of the retina, resulted in elongation of the chicks' eyeballs.

Eyeglasses as aids for defective and aging vision were invented in Italy at the end of the thirteenth century, probably by a glassblower named Salvina Armato. Armato had damaged his own vision while experimenting with light refraction and had turned to glassmaking to make correcting lenses to improve his sight. The innovation apparently caught on quickly. For many people, eyeglasses must have helped to extend working life into old age. And with the proliferation of printed material—reading—eyeglasses became a necessity rather than a luxury.

In Venice, the center of European glassmaking, craftsmen busily turned out disks for the eyes. The early lenses were convex and helped only the farsighted, those who had trouble seeing close up. The first users, naturally, were scholars, nobility, and the clergy. More than 100 years later, concave lenses, which improved distant vision and were not considered as important for intellectual pursuits, were made for the nearsighted. They became known as glass lentils because of their resemblance to the seeds of the popular legume, and eventually as lenses. Individuals peered through various lenses stocked in a craftsman's shop and chose those that best improved their vision. Accurate glasses made to correct specific flaws in an individual's vision had to await such developments as Hermann von Helmholtz's 1851 invention of the ophthalmoscope for measuring the defect and modern techniques for mathematically exact lens polishing.

A famous early eyeglass wearer was Giovanni Cardinal de'Medici, who became Pope Leo X in 1513. He liked hunting and had several pairs of eyeglasses with concave lenses to aid his marksmanship. Four years after he became pope, he sat for a portrait by Raphael wearing his correcting lenses.

In the early centuries of their use, eyeglasses were a status symbol; later, when more people used them, they were considered unfashionable, especially for women, and were worn only in private. The earliest eyeglasses either sat precariously on the nose or were kept on by leather straps tied behind the head or small circles of cord that looped over each ear. Today's rigid-sided temples, which hold the lenses more firmly in place, were invented in eighteenth-century London.

Benjamin Franklin (1706–1790) designed bifocal lenses so that he could glance up from reading and enjoy the scenery without having to switch eyeglasses. Today we have trifocals as well. Today's eyeglasses, made of plastic, are lightweight and less tiresome for the wearer, but historically eyeglasses have been heavy, with frames made of ivory, bone, tortoiseshell, or metal.

Sunglasses, the first ones consisting of lenses tinted by smoke, were invented in China but not as aids to vision. Chinese judges wore them so their expressions could not be seen, keeping their evaluation of evidence secret until the end of a trial. Modern sunglasses were developed in the 1930s by Bausch & Lomb for the Army Air Corps to protect pilots from high-altitude glare. They were tinted dark green to absorb light in the yellow band of the spectrum, and the drooping shape shielded the pilots' eyes as they glanced down.

Leonardo da Vinci (1452–1519), who seems to have thought of everything before its time, proposed contact lenses in his *Code on the Eye*. Da Vinci's scheme was to correct vision by placing the eye against a short, water-filled tube sealed at the end with a flat lens. The water came in contact with the eyeball and bent the light rays like a curved lens.

The first contact lenses were developed in 1877 by a Swiss physician, who proved that vision could be corrected with direct application of a lens and that the eye could learn to tolerate a foreign object of glass. The glass was blown or molded to the appropriate curvature, polished, then cut into a lens that covered the entire exposed eyeball. In the 1930s the German firm of I. G. Farben introduced Plexiglas lenses, and in the 1940s American opticians produced corneal lenses that covered only the central portion of the eye.

Today a person who needs vision correction can choose among hard contact lenses, gas-permeable hard lenses, and soft contact lenses. Each has its strong points. Hard lenses are better at correcting astigmatism, give clearer vision, and are easier to clean of protein deposits, oil, and dirt. Soft contacts, mostly water, are more comfortable next to the eye but are not as durable and require more care to keep clean. They resemble little pieces of plastic wrap. Both kinds of lenses are available in a rainbow of colors; we can now all have Elizabeth Taylor's violet eyes, or at least the color. Disposable contact lenses are on the horizon but will be more expensive. One of the worries about them is that people will wear them for a longer time than they really should, increasing eye problems.

Extended-wear lenses are also available, but with extended wear comes greater risk of corneal ulcers and infections by acanthamoeba. My ophthalmologist thinks a week of wear at a time with extended-wear lenses is plenty. "Would you wear your socks that long?" he asks.

Another possibility is *radial keratotomy*, a surgical procedure in which radiating cuts are made in the cornea. It results in a flatter cornea and can correct nearsightedness with 90 percent success. That success rate isn't enough for my fussy ophthalmologist, however, and he doesn't perform the procedure, which he regards as crude. "You can't control it enough," he says. For the 10 percent who become farsighted from the procedure, it's disconcerting, and it's difficult to wear contact lenses over a cornea that has been surgically altered by radial keratotomy.

Vision Loss

Like the auditory apparatus, the visual apparatus can be afflicted or damaged by a variety of things.

I knew two blind brothers when I was growing up, who were anything but subdued in their activities. They liked to ride bicycles. They were born blind, but no one knew why. In Arizona, I met a blind college student whose retinas were detached by a football injury.

Imagine the retina as a balloon blown up inside a tennis ball. A hole is poked in it. Deflated, it pulls away. Surgeons can reinflate the retina indirectly, using injected air or gas, but it must be treated right away. The delicate retina cannot be touched by a suture or scalpel. Some cases of partial retinal detachment have been related to stress. In *acute chorioretinopathy*, the retina detaches at the area of most acute vision. The problem occurs most often in men ages 30 to 50 who have had a distressing psychological experience, and in most cases the retina repairs itself in several weeks or months. Blood pressure changes are thought to be the underlying physical cause.

My great-uncle Leander was blinded as a baby by spinal meningitis. He was a musician and a piano tuner and had worked as a telephone switchboard operator when he was younger. I remember him as a handsome, gray-haired man, with wandering milky blue eyes and long fingers that were always moving. When I was a child, those fluttering fingers touched my head and shoulders to assess how I had grown, and he always ran his hands lightly over my slick braids or my ponytail and told me I looked pretty. I thought that was funny. How could he know?

Dennis, a blind engineer at my university radio station, was a victim of childhood *glaucoma*, a plumbing problem in which the aqueous humor does not drain properly. Fluid in the anterior chamber presses on the posterior chamber of the eye, causing permanent damage to blood vessels and neurons of the retina. Dennis tells me that he lost his color vision at age eight and could no longer distinguish light and dark by the time he was 12. He has two glass eyes, and he met his seeing wife on a blind date. He winds down the radio station corridors easily, his hands reminiscent of Uncle Leander's, lightly touching the walls. He produces and records radio programs. He likes to fix radios, appliances, and cars, but the miniaturization of parts has made repairs much more difficult, he says. And probably not just for him.

Today, ophthalmologists routinely test for glaucoma, measuring the pressure within the eyeball via a machine called a *tonometer*, which puffs air onto the eyeball. But sometimes people have lost 50 percent of their optic nerve before the damage caused by glaucoma is picked up on a visual field test. Eight percent of Americans over 40 have elevated pressure in their eyes, and 20 percent of them go on to develop blindness. Treatments for glaucoma vary: Drugs can increase outflow or decrease inflow of fluid. A laser beam, used like a spot welder on the trabecular meshwork where the iris meets the cornea, can open up the drains. An external exit can be slit between the conjunctiva and sclera to form a bleb, a little resorption pool. Some people become blind in spite of all treatments, however, and some become blind even with normal fluid pressure.

Diabetic retinopathy is also notorious for blinding. In diabetics, abnormal blood vessels can grow in the eye, seemingly in areas that are not oxygenated well. (In one group newly diagnosed as diabetic,

18 percent had retinopathy.) The blood vessel growth seems to be a response to oxygen starvation, but the vessels are fragile and brittle and easily ruptured, and it is leakage of blood that causes blindness. In some cases, the vitreous humor clouded with blood may be removed and replaced with saline solution. Or that spot welder, the laser, can be used as a preventive measure, ablating the retinal areas screaming for oxygen and causing atrophy of the abnormal blood vessels.

Abnormal vessel growth in the eye also occurs frequently in premature babies and can result in vision loss. This retinopathy of prematurity has been treated successfully by cryotherapy, applying freezing cold to the white of the eye to stop vessel growth.

Age takes its own toll on the eyes. By about age 65, pupils have become sluggish. In dim light, they don't dilate as widely. Because the retina receives only one-half to one-tenth as much light as younger retinas, older people with this *senile miosis* may have diminished dusk or nighttime vision and thus be reluctant to go out at night.

Macular degeneration is the most common cause of poor vision in older people, causing blurring, blind spots, and lack of contrast sensitivity. When my ophthalmologist looks inside an eye with macular degeneration, he sees what looks like dirt on the retina, or specks of pigment like pepper. Oral administration of zinc seems beneficial in stabilizing macular degeneration, he says, but will not reverse it.

When the transparent lens becomes opaque and unable to admit light, it has become a *cataract*. Cataracts may result from trauma or disease but are most commonly a result of aging and are formed when proteins become stuck together inside the long cells of the lens. Steroid hormones and damage from ultraviolet light may contribute to both glaucoma and cataracts. Fifteen percent of Americans aged 75 to 79 have cataracts, for which the only treatment is removal of the lens. Most often, artificial lenses are implanted, but contact lenses or thick glasses may be used to do the light-bending job the lens was doing earlier. Cataract removal was done as early as 1000 B.C. in India. If the operation was successful, the surgeon was paid well. If a wealthy freeman lost his vision as a result of the surgery, the surgeon's hand was cut off. If a slave was blinded, the surgeon was obliged to replace him.

Cornea damage can also fog the eye's window. The damaged portion can be removed and a new piece patched in. The first corneal transplant was done in the late nineteenth century, and today the operation is commonplace. Tissue typing is unnecessary because the cornea has no blood supply carrying antibodies to attack the foreign tissue.

The accessory structures of the eye have their own problems with microorganisms. Pinkeye is an inflammation of the conjunctiva, the mucous membrane on the inside of the lids that also bends down and covers the front of the eye. Pinkeye is caused by infectious agents—bacteria, fungi, protozoans—or by irritating substances. A sty is a little abscess caused by a bacterial infection around an eyelash hair follicle. *Trachoma* is an infectious bacterial disease of the conjunctiva and cornea.

Smaller and Larger

The development of microscopes, extending human vision into the world of the small, and telescopes, extending vision outward from Earth, has created new fields of science and more and more things for humans to think about, conceptualize, and name. Think of it: cell biology, planetary science—all because of lenses and mirrors.

The optical properties of curved surfaces were known by 300 B.C., but the microscope came 2,000 years later. The invention of the compound microscope, with two sets of lenses (not so different from microscopes used today in biology classes), is credited to the Janssen brothers, Dutch lens grinders, in 1590. More famous for his lenses and microscopic observations, however, was the Dutchman Antoni van Leeuwenhoek (1632–1723), a dry-goods salesman and janitor who ground lenses as a hobby. He fixed his tiny lenses in a flat piece of metal, equipped with a handle and a needle sticking up behind the lens. Upon the needle he placed all sorts of things—scrapings from teeth, scummy water, semen. He described and drew pictures of the little cavorting "animalcules" that he saw and reported his observations to the Royal Society. Another early microscopist was the Englishman Robert Hooke (1635–1703), who presented the results of his investigations of cork to the Royal Society. He used his penknife to slice the cork into thin sections that light would pass through, and he called the little boxes he saw "cells." Hooke was actually observing the cellulose cell walls of dead and empty plant cells. By 1839 scientists had seen enough cells to propose a cell theory, the idea that all living things are made up of cells, and by 1858 they had seen enough cell division to be confident that all living cells arose from preexisting cells, a crucial advance in biology.

The light microscope increases the apparent size of the object being viewed until it provides an adequate stimulus to our eyes. But its powers are not infinite. The resolving power of the light microscope—its capacity to separate adjacent objects—is limited by the diffraction of light. The resolving power of a light microscope is about 500 times better than that of the human eye.

In the 1930s the electron microscope was developed, opening up new vistas in the study of cells. The electron microscope has a vacuum chamber and uses a beam of electrons instead of light as a source of illumination. It can resolve objects 500,000 times better than the human eye. With the electron microscope, the magnification can be equal to increasing the length of a human hand to 25 miles long. Transmission electron microscopes (TEMs) enable looking at very thin sections of material, and scanning electron microscopes (SEMs) allow looking at whole specimens (giant insect pictures are taken with SEMs). Preparing specimens for the electron microscope is a lot of work. They must be impregnated with heavy metals or coated with metals such as gold palladium. The microscope works by stimulating the emission of secondary electrons from the metal, so what we see is not the real thing but an artifact of the real thing.

Astronomical telescopes were also an outgrowth of work with lenses. Galileo (1564–1642) is credited with building the first telescope by putting two lenses together. Today two kinds of optical telescopes are used (as well as radio telescopes, which pick up radio waves that are then converted into pictures): a refracting telescope, like Galileo's, which uses lenses, and a reflecting telescope, which uses mirrors. Binoculars consist of two refracting telescopes mounted side by side. The big research telescopes, however, are reflecting telescopes. A concave mirror at the bottom of the tube reflects light back out to a focus point, and various arrangements pick up the focus. The larger the telescope, the more light can be gathered from space, thus more detail. The weight of the key element, the mirror, has been a limiting factor in telescope size. Around the world, there are 28 telescopes with apertures exceeding two meters. The Soviet Union currently has the largest reflector telescope, with a six-meter mirror. The Hubbel telescope, designed to be launched from the space shuttle, has a 2.4-meter mirror. But its capacity to gather information about our universe will be enhanced by its position in space, away from the clutter of Earth's atmosphere, and it is expected to revolutionize astronomy.

Tears

Every few seconds during waking hours, our eyelids blink briefly, wiping tears over the eye. The tears are secreted continuously from lacrimal glands located above the outer corners of each eyeball. Continuous tears consist of water, mucoid proteins, and lysozyme, an enzyme that destroys bacteria by digesting them. Excess tears drain into the nasal cavity through the nasolacrimal duct, at the inner corner of the eye, and are swallowed. Irritating substances, such as onion volatiles or ammonia, will cause greater tear secretion, as will bright light, irritation of the eye, and certain drugs. The eyes also water or tear when a cold or allergies block the drainage of tears into the nasolacrimal duct. Sniffles associated with tears result from an excessive amount of lacrimal secretions draining into the sinuses.

But humans also produce emotional tears, tears associated primarily with grief and joy, and they have been found to have a different chemical makeup than tears of lubrication or irritation. Biochemist William H. Frey II and his colleagues have done extensive studies of psychogenic tearing (and Frey has written a charming account of how they went about these unusual studies in *Crying: The Mystery of Tears*). They found that the protein concentration of emotional tears was 21 percent higher than that of irritant tears. Frey has suggested that emotional tearing may play a vital role in maintaining body homeostasis by removing harmful substances. He proposed that the reason people feel better after crying is that tears may remove chemicals that build up as a result of emotional stress and that the study of tears will lead to a better understanding of the biochemistry of emotion.

Certain hormones seem to exert a kind of permissive effect on tears. Women cry more than men, an average of 5.3 times per month, compared with 1.4 times per month for a man. The average length of a crying episode, for both sexes, is six minutes. Forty percent of the crying episodes reported by women are triggered by interpersonal relations, as are 36 percent of male crying episodes. Sadness accounted for half of all crying; a fifth of the female episodes were associated with happiness. People reported crying triggered by myriad emotional experiences, including sympathy, anger, anxiety, fear, music, church services, physical pain, and weariness.

Babies wail and produce tears of irritation but don't begin to produce emotional tears until they're several weeks to three months old. An average one-year-old, however, cries about 65 times a month. Some have suggested that crying may have originated as a kind of infant "lost" call that has been elaborated in humans. Although there are many anecdotal accounts of various kinds of animals shedding tears, it seems to be the exception rather than the rule (and is a difficult subject to study). Emotional tears seem to be particularly human.

Dilation with Pleasure

Reflexes of the pupillary muscles determine the amount of light admitted into the eye, but pupil response has also been intimately linked with mental activity. The eyes are truly windows of the soul, connected to emotional states. We spend a great deal of time looking into others' eyes as we communicate with them. Eye-to-eye contact tells us many things that we may not be aware of on a conscious level. The psychologist Eckhard Hess recounts a story of how, in the early 1960s, he was lying in bed looking at a book of beautiful animal photographs. His wife insisted the light must be bad because his pupils were unusually large. He was puzzled because the light seemed adequate. He recalled that someone had once reported a correlation between pupil size and emotional response. He was intrigued. He began doing experiments.

Controlling for brightness, he found that pupil size varies with the interest value of a visual stimulus. Widening of the pupil indicates a positive response; constricting a negative response. The male and female subjects Hess tested differed significantly in their responses to certain pictures. Females showed positive responses to pictures of babies, to pictures of mothers with babies, and to male pinup figures. Males responded positively to female pinup figures. The pupils of males widened when they saw pictures of sharks, the pupils of females constricted when they saw the sharks. Aggression vs. fright? Who knows?

Hess also showed males two photographs of a woman, identical except that one had been retouched to make the pupils larger. The males found the large-pupiled woman more attractive, although they were unable to say why. The pupils of babies have also been shown to dilate more to pictures of human faces than to geometric shapes and to dilate more to pictures of their mothers than to pictures of strangers. Some of the people in Hess's study who insisted they liked modern art showed strong negative responses, as measured by pupil response, to almost all the modern paintings they were shown.

As long ago as the Middle Ages, women dilated their pupils with the drug belladonna ("beautiful woman" in Italian), perhaps to convince men they found them overwhelmingly interesting. Centuries ago, it has been said, Chinese jade salesmen watched their customers' eyes for dilation, indicating interest in purchasing their wares.

Pupils respond not only to visual stimuli but also to stimuli affecting the other senses. The pupils become larger when music is being played. Studies also indicate that pupil dilation may indicate taste differences too small for a subject to consciously notice and that pupil dilation can be used as a measure of the arousal value of a stimulus.

Old Eyebrow Greetings and New Blink Science

Our eyes are a focus of communication with others of our species. Among the human facial expressions observed by the behavioral biologist Irenaus Eibl-Eibesfeldt is the raised eyebrow greeting—the eyebrow flash. In a friendly greeting over a distance, the eyebrow flash is transmitted in combination with smiling, head tossing, and nodding. Watch for it when you see someone who greets you. It's a universal and stereotyped pattern among humans. In 155 cases of quick eyebrow raising, he distinguished these categories: 45 for greeting and parting, 9 for flirting, 12 for women joking with babies, 4 for thanks, 31 for affirmation, 14 for agreeing with a statement, and 40 for emphasizing an assertion.

A blink is not just a blink, either, scientists have found. We blink much more often than is necessary for eye maintenance. Psychophysiologist John Stern says blinks can indicate anxiety, fatigue, boredom, or activities such as storing information, making decisions, performing a difficult task, or shifting visual attention.

Stern became interested in blinking while watching the Watergate hearings on television. He noticed that President Richard Nixon's blink rate increased markedly when he was asked a question he was not prepared to answer. "Nixon's speech was well controlled and he did not manifest other symptoms of anxiety, but you could see it in his eyes," Stern says. "Most politicians have learned to disguise feelings except in ways they cannot inhibit." (Television announcers, however, are trained to inhibit blinks because they are distracting to viewers.)

Among the findings of blink science: People don't blink when they are listening attentively. When they start thinking about a question, or as concentration shuts off, they blink. In problem-solving, people blink as each point of the problem is solved and presumably stored mentally. People driving in traffic blink less frequently, and eyelid closure time is short. Drivers checking their speed blink as they shift their eyes to the speedometer, and blink again as they register the information. But if a police car is following, they don't blink.

Airplane pilots blink less frequently than copilots. When they exchange roles, the blinking patterns reverse. When people are tired, blinks are less crisp, as measured by a blink wave form, like an electrocardiogram.

Because blinking can be used to detect fatigue or boredom, the National Aeronautics and Space Administration, the U.S. Air Force, the Federal Aviation Administration, and the Automobile Club of America are interested in the potential of blink science for monitoring the attentiveness of pilots or drivers.

Vision, Television

Television, that radio with a picture, became part of American popular culture in the 1950s. My small town in mountainous southern Utah was about a decade behind—we got a relay station in 1962. My senior year, the winter of 1962–63, was a winter lost in the adventures of Ben Casey, Stoney Burke, and commercials.

And that was black-and-white, with snow. We have excellent color now and bigger screens (and smaller screens, if you want one to carry with you). Average daily watching time is routinely placed at about six hours per person.

We primates dearly love our visual stimuli. In living color. Physicist and former astronaut Don Lind, who flew with the *Skylab-3* mission, tells me that after food, the most important item on board the spacecraft was binoculars. The astronauts did not take a television set, but they amused themselves for hours by looking back at Earth with binoculars.

A very ancient and fish-like smell.

—WILLIAM SHAKESPEARE

That which we call a rose
By any other name would smell as sweet.

—WILLIAM SHAKESPEARE

Give me an ounce of civet, good apothecary, to sweeten my imagination.

—WILLIAM SHAKESPEARE

Smells are surer than sights or sounds to make your heartstrings crack.

—RUDYARD KIPLING

Chemical signals might serve the function of global hormones, keeping balance and symmetry in the operation of various interrelated working parts, informing tissues in the vegetation of the Alps about the state of eels in the Sargasso Sea.

—LEWIS THOMAS

Go outside and blow the stink off you.

—MY MOTHER

F · O · U · R

▼

Smell: The Inarticulate Sense

SMELL (connected etymologically with "smoulder" and "smoke"), a sensation excited by the contact with the olfactory region of certain substances, usually in a gaseous condition and necessarily in a state of fine subdivision. The sense is widely distributed throughout the animal kingdom.

Fishing for Chemicals

In his novel *Perfume: The Story of a Murderer*, Patrick Süskind creates an unusual monster: Grenouille, a man who has no odor of his own but whose sense of smell is so well developed that it is his sole motivation. Grenouille is a personification of olfaction. Born to squalor in stinking eigthteenth-century France, he becomes a consummate perfumer, his nose performing chemical analyses better than a modern spectroscope. As he follows the trails of odors, as he extracts the essence of flowers and virgins, his adventures take stranger and more fantastical turns.

Perfume turned me on to smell. For days after I read of Grenouille's olfaction, I noted comments about smell. I watched dogs nosing close to the ground. I sniffed Cruiser the cat's sniffing and rubbing spots and smelled nothing. I surreptitiously sniffed people in crowds, my friends. I sniffed myself. I squirted perfume samples at department stores and wondered at their formulation and at my likes and dislikes. I became aware of my nose's contribution to my experience and of its limitations.

The sense of smell is a chemical sense. Living things are made up of chemicals. In general, the sense of smell gives us a sense of other living or organic things, of our own species, and of other animals and plants. It plays important roles in sex and hunger, forces that rule the living world.

Perception of the chemical environment is a basic necessity for animals. It is conceivable, writes the biologist Edward O. Wilson, that somewhere on other worlds are civilizations that communicate entirely by the exchange of chemical substances. But chemical communication and smell are not insignificant here on Earth. The social insects, with chemicals that communicate about colony membership, sex, and other useful things such as trails to food, provide a close approximation to that now. Every kind of vertebrate animal has a chemical sense, testimony to its widespread usefulness. In contrast, some vertebrates have lost the ability to perceive sound, and some have lost the ability to perceive light.

In humans, the sense of smell seems an ethereal, mystical sense, transcendent as perfumes, capable of evoking nebulous ideas and forgotten scenes as well as strong, sudden emotions. Odors play a subliminal, subtle role in our lives. Humans have a strong visual and auditory point of view, and in many ways we have found it easier to analyze questions from visual and auditory perspectives. Only in recent decades has the sense of smell come under the close scrutiny of biologists, animal behaviorists, and psychologists, with some especially intriguing findings about its role in sexual contexts. Much is still to be revealed.

The senses of vision and hearing are guarded by membranes, fluid baths, and bones, and the signal is modified as it is transmitted. The sense of smell is anatomically simpler. The olfactory receptor cells protrude into the environment directly from the olfactory bulb of the brain, guarded only by a layer of mucus. They encounter an ever-changing stream of molecules containing complex information, and they make sense of it. The sense of smell has fewer limitations than the senses of vision and hearing. For vision, light and a clear path are necessary to transmit the image, making vision of limited use at night. Sounds are good for nocturnal communication, but again open environments are important for sound transmission. Sounds can be distorted and deflected by obstacles. Olfaction, however, requires only chemicals traveling up the nose. Signals for vision and hearing are immediate. Chemical signals linger.

In the human brain, the small paired olfactory lobes or bulbs, consisting of perhaps 20 million cells, are cradled in an area of the skull where the two frontal bones of the skull come together like cupped hands. Between the frontals, a flange—the "cock's comb"—of another bone, the *ethmoid*, intrudes. The ethmoid is light and fragile as a sponge, and the part on which the olfactory lobes sit is the *cribriform plate*, pierced with holes like the top of a salt shaker.

Olfactory receptor cells are located below, in the epithelial or covering layer of cells of the upper nasal cavity. Each receptor cell has a projection that reaches down like a fishing line into the layer of mucus that lines the nasal cavity. Branching from the projections are hairlike cilia, increasing the surface area for olfactory reception. In the other direction, the receptor cells reach through the holes in the cribriform plate, connecting directly with cells of the olfactory bulb, which can then send messages to many other parts of the brain, including the hippocampus, which has been implicated as an intermediary in memory. Smell, with its minimal switching, is the sense with the most direct connection to the brain. Unlike the case for vision or hearing, no one brain-receiving area has been associated with smell. In fact, the most highly developed brain centers in mammals arise over the area directly connected with smell— smell may be distributed all over the brain.

The *nasal septum* divides the nasal cavity into two spaces, each folded into scrolls, the *conchae*, which increase surface area and cause the entering air to eddy and swirl. Because the chemicals we detect must be in solution, a remnant of our watery beginning, our nasal passages and those of the other terrestrial vertebrates are moistened by a constant production of mucus, about a quart a day. Fishes and aquatic amphibians encounter chemicals that are already in solution; consequently they have no mucous membrane and never have runny noses.

The air we breathe in from the atmosphere is about 79 percent nitrogen and about 21 percent oxygen, with a fraction of a percent of other gases. The air also contains suspended particulate matter—dust—along with a variety of chemical molecules that give us clues about the external environment. As we breathe in, nostril hairs catch the large dust particles. As the air swirls through the conchae, it is warmed and moistened for a more gentle entry into the windpipe. The air also passes by postage-stamp-size islands of yellow-brown olfactory epithelium on the upper conchae and septum. The odorant molecules contained in the entering air are dissolved in the mucus, diffuse through to the hairlike cilia of the receptor cells, and stimulate them. Many theories have been proposed, but the exact method of stimulation is unknown.

The olfactory receptors can detect and discriminate thousands of different airborne chemicals. Our sense of smell is 10,000 times

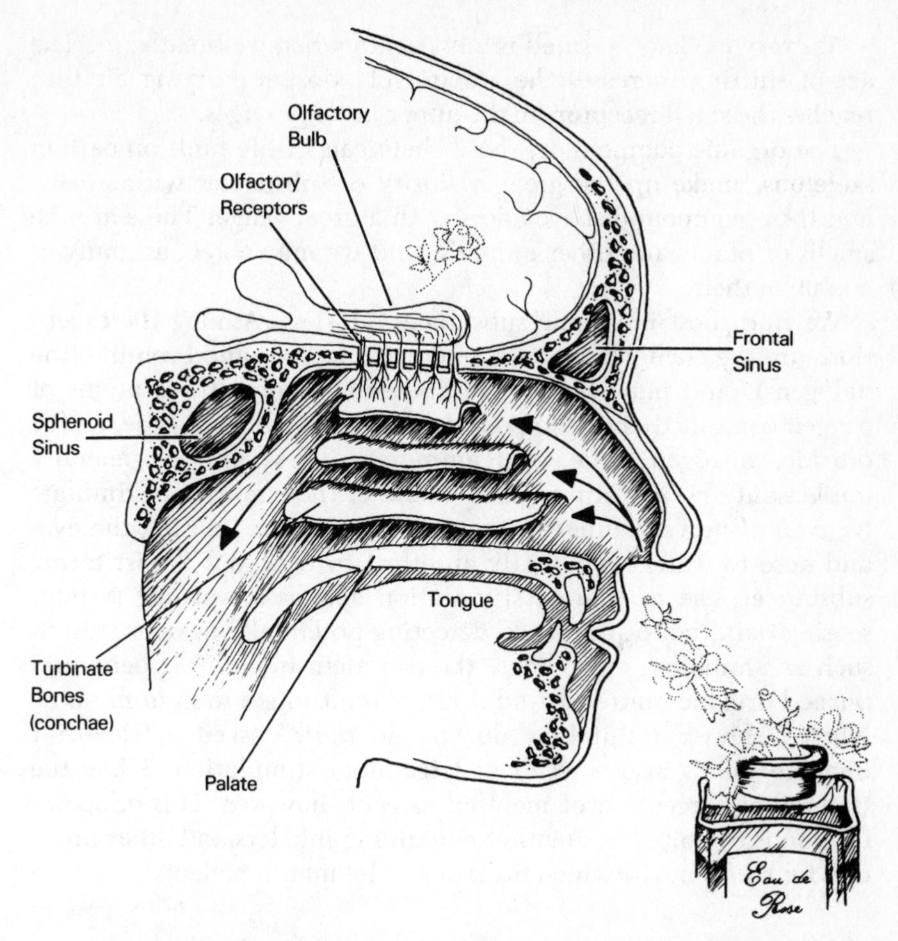

The human olfactory apparatus, a sophisticated chemical detector, is located high in the nasal passages. The "fairies" may be protein carrier molecules produced near the tip of the nose.

more sensitive than our sense of taste (another reason, in addition to its inconvenient location, that olfaction has been difficult to study, and understanding of smell in 1940 was said to be 200 years behind understanding of hearing and vision). For a substance to be smelled, it must be volatile; that is, it must enter a gaseous state at ordinary temperatures so that it can reach the upper airways. It must also dissolve in the mucus and reach the olfactory cells. When a person has a cold and the nasal cavities are coated with heavy mucus, the sense of smell is handicapped (and food is tasteless).

As everyone knows, smell is most acute when we breathe in. The act of sniffing increases the amount of odorant-carrying air that reaches the smell receptors in the upper nasal passages.

The organic compounds, those chemicals of life built on carbon skeletons, make up the great majority of substances with smells, and these compounds have odors with a great range. These are the smells of plants and other animals, and we may detect as many as 40,000 of them.

We find most inorganic substances odorless. Among the exceptions are the elements fluorine, chlorine, iodine, and bromine (the halogens), and phosphorus. We also smell ozone (three atoms of oxygen combined) and compounds such as hydrogen sulfide, sulfur dioxide, nitrogen oxide, and ammonia. These are all generally unpleasant and irritating smells. Some of these may also stimulate *trigeminal* nerve endings in the respiratory area, causing the eyes and nose to water, apparently an effort to wash out the irritating substance. The trigeminal stimulation acts as a warning system, sensing pain, hot or cold, and detecting potentially noxious stimuli such as ammonia, cut onions, the capsaicin in chili peppers, and burned organic matter. I find it unpleasant to get soap in my nose when I shower ("But how do you do *that*?" asked a friend); I imagine this is also a result of trigeminal stimulation. I like the trigeminal perception of menthol as cool, however. This olfactory illusion is exploited in menthol-containing inhalers and other products for stuffy noses (which my sister calls "man-repellents").

An Ancient Sense

Cells that detect chemicals are found in the lower invertebrates, primitive animals without backbones, and the chemosenses are especially well developed in the insects, highly evolved invertebrates. They have chemoreceptors on mouths, legs, antennae, and female ovipositors.

A kind of olfactory cell is found in *Amphioxus*, a primitive fishlike animal thought to be similar to the ancestors of the vertebrates. In fishes, the nasal structures with their sensory cells consist of a pair of pockets in the head with fore and aft external openings so that water constantly flows through over what are sometimes immense areas of folded sensory tissue. The nasal structures do not open into the mouth; that arrangement came later, with the amphibians. In the history of vertebrate animals, nasal association with breathing is a secondary function.

Among the vertebrates in general, through bony fishes, amphibians, reptiles, and most mammals, smell provides a main source of information about the environment. Through chemical memories, salmon seem to find the waters of their origin for spawning. Smell brings the ridley sea turtles back to the islands where they were hatched to lay eggs. A rat remembers a food that previously made it sick and avoids it. A wolf identifies territory marked by another wolf. A human remembers a tasty food, a love-filled night.

The chemosensory apparatus of a salmon is in the top of its head, separate from the mouth. It allows constant analysis of water passing through. Salmon remember their birthplaces and find their way back from the ocean to spawn in small continental streams where they were hatched. Atlantic salmon return year after year; Pacific salmon die after spawning. This homing instinct, which takes a salmon hundreds of miles, seems to be related to a smell memory of odors from its parent stream.

Dogs have a larger olfactory surface than do humans, and can identify individuals by their odors.

Chemicals can transmit a large amount of information efficiently, and chemicals are cheap in terms of the energy required to produce them. Chemicals in low concentration can have a tremendous effect on organisms. Through chemicals, animals can recognize food (or nonfood), mark territories, repel predators, and identify enemies, friends, and mates. Chemical signals provide information about species, sex, and individual identity, as well as information about dominance, reproductive status, dietary status, and health.

In most vertebrate animals, the nasal regions occupy up to half the length of the skull; some of the enlargement may be for the warming and moistening of air, but the increased area provides greatly increased olfactory surface as well. In general, the carnivores and the rodents have the largest olfactory areas. When my dog friend Sal leaves the house in the morning and inhales, he exposes approximately 40 million olfactory receptors in each square centimeter of his olfactory epithelium to the organic molecules carried on the morning breeze—shades of Grenouille's perceptual world. (Cats' noses are sensitive too. Cruiser was in the hospital once when Sal came to visit. Upon his return, Cruiser ran around the house and sniffed everywhere Sal had been.)

Dogs, lions, and wolves locate prey by smell, and their tracking abilities are well known. They identify individual humans as well as others of their species by chemical signatures. "Stand, smell, and be smelled," is the standard procedure when Edie and I are walking with Sal and he meets another canine. Diane's old dog, Skeeter, whose eyes and ears are failing, walks to the door to greet every visitor with his nose, about calf-high. Satisfied with the smell-print, Skeeter lies back down.

Rodents, on the other hand, are not trackers. They eat plants or seeds, items that stay put. Their sense of smell seems to be most critical in social contexts, in recognizing other rodents individually and in discriminating among families or tribes.

Some mammals, reptiles, and amphibians also have a specialized olfactory structure, the *vomeronasal organ*, or Jacobson's organ,

The gila monster, a large poisonous lizard of the American Southwest, "smells" the air by gathering airborne molecules on its forked tongue, then inserting the tongue tips into a chemical-sensitive structure in the roof of its mouth. Other snakes and lizards flick their forked tongues for the same reason.

separate from the olfactory tissue in the main part of the nose. In some cases it may pick up olfactory sensations from food in the mouth. In mice it seems to play a role in aggressive and reproductive behavior. In lizards and snakes it has been elaborated into two pockets in the roof of the mouth. These reptiles flick out their two-pronged tongues to pick up chemicals from the air, then draw in their tongues and insert the tips into the sensory pockets to smell the air they have sampled. The vomeronasal organ is also present in newborn humans but shrinks with age.

In vertebrate groups where the eyes have become the dominant sense organ, olfaction is less well developed. This is true of most birds, for instance, and of the primates, the group to which we belong. Olfaction is also minimal in bats and in the toothed whales. Insect-eating bats depend on echolocation, a kind of sonar system, for finding prey. Whales, air-breathing mammals descended from terrestrial animals that returned to the sea (protein studies indicate they are related to pigs), apparently were not able to readapt their terrestrial nose design to the aquatic environment. For them, hearing has become the sense most highly developed and extremely important in communication.

Flying animals, the birds and bats, certainly have less need for an acute sense of smell than animals that spend their time on the ground. But my friend Bob Parmenter, who spends his time in the strange scientific study of life after death—the dynamics of decomposition and the recycling of organic matter—tells me the turkey vulture hunts by smell in tropical rain forests of the Americas. It has large olfactory bulbs, compared with those of other birds, and big external nasal openings. The turkey vulture is also efficient in locating day-old carcasses in the jungle. In a study in which dead chickens were placed on the forest floor in Panama, the vultures removed about 90 percent of them within a very short period of time.

Dead animals don't last long in tropical forests. Ground scavengers, especially ants and carrion beetles, are always scouting for carcasses, and the heat and humidity provide ideal conditions for the rapid growth of ever-present bacterial decomposers. For a turkey vulture to compete for the carcass, it needs to get there quickly. The vulture apparently cues in on clues provided by early bacterial activity; that is, the chemical metabolites produced as bacteria begin their work on the carcass. The bacterial toxins and unpalatable byproducts may deter other scavengers, but turkey vultures have botulism-resistant nervous systems and heavy-duty immune systems to ward off poisoning. In one experiment, a scientist injected a turkey vulture with enough botulism toxin to kill 300,000 guinea pigs. The bird was unperturbed. (The vulture's naked head, exposed to bacteriocidal ultraviolet light, is another adaptation for its corpse-eating life-style.) Contrary to popular belief, however, vultures avoid badly decayed corpses, avoiding assaulting their systems with too many toxins. Because the chemicals given off by rotting meat vary with time, as bacterial populations change, vultures probably use their sense of smell to distinguish the relative freshness of a carcass.

When a stinky sulfur-containing mercaptan, ethanethiol, is injected into gas lines, some will escape from leaks. In 1938 a Texas oil company engineer discovered that turkey vultures will circle around

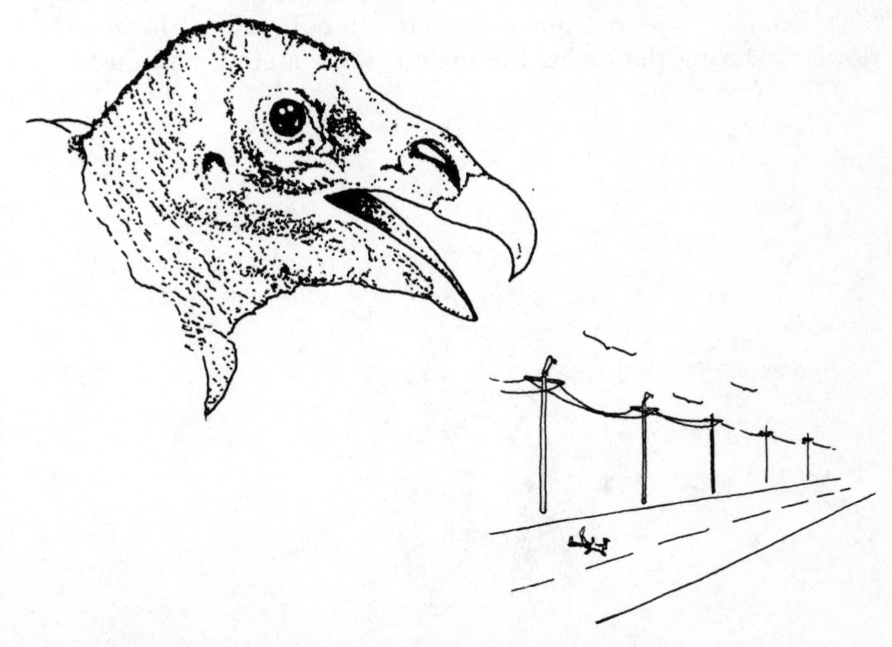

Turkey vultures of the Western Hemisphere have large nostrils and a good sense of smell, which will lead them to fresh corpses in the jungle as well as to leaks in natural gas pipelines.

the rising plume of escaping mercaptan and descend nearby. A gathering of turkey vultures around a gas line signaled a leak as surely as expensive leak-detecting equipment.

Carrion and fecal smells, also unattractive to us, have the opposite effect on certain flies and beetles that find in their source a place to lay eggs and food for their developing larvae. Some plants have capitalized on this characteristic and use it for their own ends: for pollination, seed dispersal, or in the case of the insectivorous plants, food. Some plants, not human favorites, are great impostors, producing chemicals that smell like rotting flesh, festering wounds, congealed blood, feces, or urine.

The European cuckoo-pints produce an odor that spreads over a wide area. The plant helps disperse the odor by heating it, developing midday temperatures up to 15 degrees Centigrade greater than that of the environment. One species is particularly attractive to a small fly, the owl midge, which frequents places of human feces. The midges land on the plant's slippery cuplike structure and slide down past rings of spines that prevent them from escaping. If they have brought pollen from elsewhere, it adheres to the female flowers. During the night, the male flowers mature and dust the

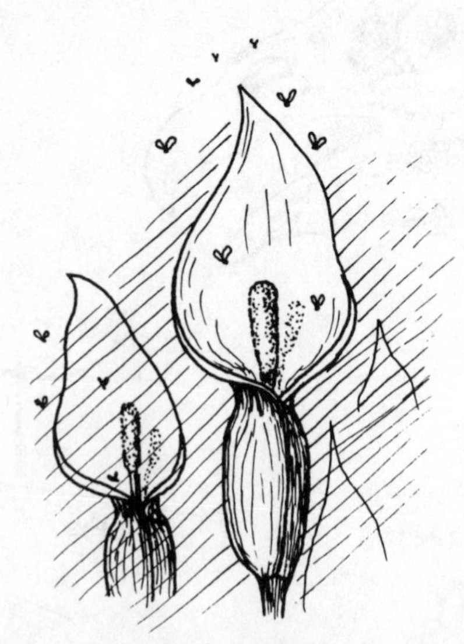

Arum maculatum, *the cuckoo-pint, is a member of a widespread group of plants that produces fetid odors attractive to carrion flies, tricking the flies into providing pollination services.*

midges with their pollen. The plant spines shrink, and the flies crawl out. Self-pollination is prevented, and the midges are free to repeat their misadventures the next day in their search for fecal smells.

The odor of the European stinkhorn, *Phallus impudicus*, has been traced to the chemical phenyl acetaldehyde, which has not been identified as a component of decaying flesh but works in the same way. (The real thing and the mimic don't have to produce the same signal as long as the signal-receiver reacts similarly to both.)

Humans are repelled by smells of decomposition and of feces, in which potentially harmful bacteria are also present. "We evolved from frugivores—fruit eaters," Bob says. "Those of us who didn't consider the smell of rotting flesh awful died of ptomaine poisoning. We survivors stay away from rotting meat."

When I studied zoology and primatology, and later when I taught zoology courses, I found that textbooks dismissed the human sense of smell as much less important than the other senses. As the primates got their noses off the ground and evolved into vertical clingers and leapers, the story goes, the doggy wet nose—the rhinarium still present in the lower primates, such as lemurs—disappeared. Then another decisive step in human development was taken, as Sigmund Freud says, when early man got his nose above the female genitalia and assumed an erect position and a two-legged gait. Vision became paramount. Bipedalism gave the human eyes a new point of view for gathering environmental information and freed the human hands to manipulate food, making the hands and the sense of touch contained in the fingers an adjunct of the gustatory and digestive systems.

The sense of smell is less acute in humans than it is in dogs, but it has not become unnecessary. True, we are not nosing the ground as substrate sniffers, and our noses are not crotch-high. The significance of smell for humans is more subtle, but it is there nonetheless—in sexual communication, as we shall see, and in food-getting contexts. Our bipedal gait changes the proximity of the crotch from the nose, but human males secrete the aromatic sexual communication chemical androstenone in their underarm sweat, saliva, and scalp oils as well as in their urine.

And everything that enters the mouth passes under the sentinel of the nose. Humans consume a wide variety of foods, each with its volatile components that are smelled. Smell contributes significantly to nutritional wisdom. In the course of evolution, selection has favored those humans with a tendency to detect and avoid, or to learn to dislike, tainted foods and water. Selection has also favored those who enjoy eating calorie-rich, protein-rich, and vitamin-rich foods. Foods that smell good to us as they are prepared or as we eat them are generally good for us nutritionally. The odors make our mouths "water." They stimulate our salivary glands, priming us for the digestive process. Distinguishing among different odors helps us to achieve a well-balanced diet. And our memory of smells plays its part as well.

Classifying Smells

Smell wafts away from us by its inarticulateness. Smells have no names, except those of the objects they come from. Consider: The eyes see colors with names, such as red, green, yellow, blue; the tongue tastes sweet, sour, bitter, salt; the ears hear vowels, consonants, and notes that have names. But smell smells Lady Esther powder, baking bread, rotten eggs, feces, roses, a skunk, a wet dog. How to classify smells?

The early natural philosophers and physicians who included olfaction in their accounts of the senses usually made only feeble attempts at classification but did write about the effects and the benefits of smell. Plato divided odors into pleasant and unpleasant. Aristotle separated sensations that arise in the mouth and sensations that arise in the nose, noting that olfaction produced indefinite kinds of sensations. If something smells good, Aristotle said, it is good for us, and if it smells bad, it is bad for us, as a general rule. Taste was simpler to deal with. Because Aristotle thought the stimuli that caused odors were similar to those for taste, he spoke of odors as having tastelike qualities: sweet, pungent, harsh, sour, and succulent. He added a fetid odor, analogous to bitter taste. This was another reference to effect: Fetid odors, he said, made inhalation offensive, whereas bitter taste made swallowing offensive.

Aristotle's student Theophrastus (320 B.C.), who had experience with perfumery and botany, proposed a sevenfold classification scheme, but the record of the actual classes has been lost. Two thousand years later, in 1752, another serious scheme was proposed by Linnaeus, history's most compulsive and accomplished classifier. Linnaeus' scheme also included seven classes: aromatic, fragrant, ambrosial (musky), alliaceous (garlicky), hircine (goaty), repulsive, and nauseous. Various other schemes were proposed between Linnaeus' time and the end of the nineteenth century. Some were ambitious exhaustive attempts; others dealt only with classification within certain groups, such as flowers, perfumes, or fungi.

The Dutch physiologist Hendrik Zwaardemaker (1857–1930) proposed an "updated" scheme based on that of Linnaeus but which also considered other schemes and included modern chemistry. His grouping included nine classes with subclasses: ethereal (fruits, resins, ethers), high-potency aromatics (camphor, cloves, lavender, aniseed, lemon, bitter almond), floral and balsamic (flowers, violet, vanilla), ambrosial (amber, musk), alliaceous (hydrogen sulfide, arsine, chlorine), empyreumatic (roast coffee, benzene, burned organic matter), caprylic (cheese, rancid fat, goat), repulsive (deadly nightshade, bedbug), and nauseating or fetid (carrion, feces). Zwaardemaker is also the inventor of the olfactometer, an instrument consisting of two tubes, one inside the other, designed for delivering a measurable amount of odorant to a subject's nose.

Hans Henning in 1916 proposed an alternative scheme, with six main classes: fragrant, ethereal or fruity, resinous, spicy, putrid, burned. Henning took the scheme a step further. He provided a model structure for the classes on a three-dimensional prism, with the six corners of the prism representing turning points in the direction of resemblance.

The Henning model showed gradual transitions from one odor to another. For a few years it generated a great deal of interest, but it could not accommodate certain smells, and there was substantial disagreement among subjects about smells. (Although generalities can be made, noses are different. Linnaeus classified the bedbug smell as repulsive, but a subject in one study reported that the bedbug volatile smelled like fresh raspberries. This also led to speculation less maligning to the bug: that the bedbug smell was a general smell of an unclean house and the bugs' leavings, rather than the bug. Only recently have bedbugs not been familiar to everyone, it seems.)

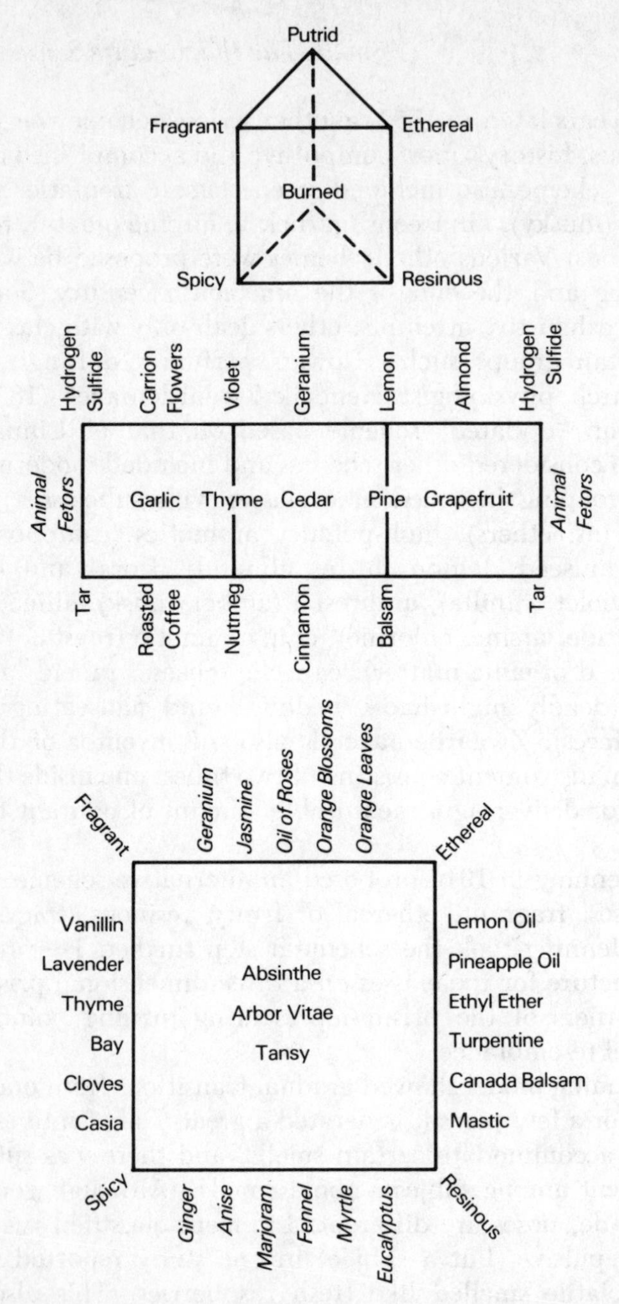

Henning's three-dimensional odor prism was an early way of trying to categorize, on its surface, odors that humans detect. (Experimental Psychology, *by R.S. Woodworth, © 1938 by Holt, Rinehart and Winston, and renewed 1966 by Svenson Woodworth and William Woodworth. Redrawn with permission from Holt, Rinehart and Winston, Inc.)*

The classification of odors, however unsatisfactory, has helped to suggest questions or hypotheses about the organization and functioning of the sensory system. For instance, do the groups reflect types of receptor cells or receptor sites? Do the chemicals in each class share physical or chemical properties? Do the classes represent primary odors that might be mixed to produce other odors?

From the first studies in which measured amounts of odorants were used, it was apparent that the threshold concentration for some odorants is extremely low. The threshold for the evil-smelling sulfur-containing mercaptans, for instance, is in the vicinity of one part per trillion. On the other hand, substances with very high or very low molecular weights are not potent odorants. Certain elements are never constituents of compounds with odor, whereas elements in the fifth, sixth, and seventh groups of the periodic table are common constituents of odorous compounds.

These discoveries led to early optimism that underlying principles could be discovered. But olfaction has been a difficult nut to crack. Theories have been proposed, tested, and discarded. In the late 1930s electrophysiological techniques were applied to olfaction, and activity in the olfactory bulb was measured. The anterior of the bulb was found to be excited by water-soluble substances, and the posterior by lipid-soluble substances. As individual receptor cells were studied, they were usually found to respond to more than one substance, and the list of substances was long. Odor quality is somehow coded by a pattern of impulses across many units of diverse sensitivity. Although the human sense of smell is not superspecialized, a trained human might detect 5,000 different odors. The actual perception of an odor depends upon the ability of neurons to compare information coming in from the activation of different receptors and to analyze and interpret it. The key may lie in a small number of receptors, actually. Differential stimulation of ten receptors, for example, could result in different olfactory profiles for hundreds of different odorants. For example, banana odorant might stimulate receptors one, three, and eight; fail to stimulate two and four; and stimulate the others to various degrees.

A functional group of an odorant molecule may react with a site on the olfactory receptors. In small molecules, at least, the functional group imparts a particular quality. For instance, low molecular weight amine chemicals are fishy and urinous; low molecular weight thiols are garlicky and sulfurous; ketones, aldehydes, and alcohols all share a group smell. Large molecules, however, may be registered by receptors sensitive to shape or fragments of shape.

A stereochemical model of olfaction proposed in the 1960s considers the three-dimensional size and shape of the molecules to be the most important factors; another, the vibrational model, con-

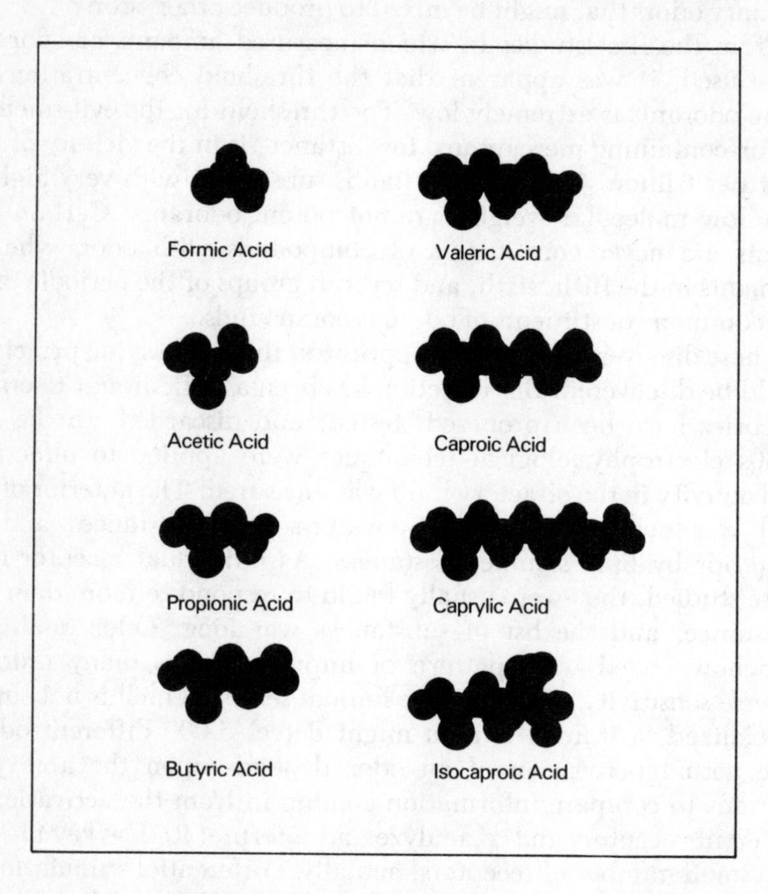

Molecules fit into "families," and in many cases similar molecular shapes confer similar odors. These are structural formulas for some pungent acids, viewed from a similar angle. Formic acid is present in the sting of ants. An arthropod known as a whipscorpion or vinegaroon sprays acetic and caprylic acid to deter predators. Propionic acid is a fermentation product of bacteria and has been used as a topical fungicide. Butyric and valeric acids are detected in rancid butter, while caproic and isocaproic acids, with "goaty" smells, are found in both plant and animal products. (From Gustation and Olfaction, *edited by G. Ohloff and A. F. Thomas, © 1971. Adapted with permission from Academic Press.)*

siders oscillations within the molecule to be of paramount importance in reception of odors. Yet another model looks at the time required for molecules of different weights to pass through the mucus. Recently, protein "carrier" molecules have also been found to be involved. Each idea may be a piece of the puzzle.

To try to learn about what is happening with olfaction on the chemical level, scientists have studied groups of related chemicals to determine what they have in common or how they differ. In the groups that include musk, amber, and woody odors, small changes in the chemistry alter the strength, but they still smell the same. In the chemicals related to the grapefruit odorant, the opposite occurs: A small change changes the odor quality but not the strength.

In the pyrazine group, which are important odorants in foods such as peanuts, cocoa, coffee, and green peppers, potency has wide variation. One scientist isolated, identified, and synthesized the pyrazine that provides the odor of green bell peppers. The odor threshold was found to be low: two parts per ten trillion parts of water. Months after the bell pepper compound was made, the smell permeated the end of the building where the scientist's laboratory was.

Scientists have found that a small change in the structure of the lactone chemicals will change a smell from coconut to peach to celery to beef bouillion. In other cases, many different groups of chemical compounds with very different structures—for example, steroids, macrocyclic ketones, nitrocyclohexanes, indanes, tetrahydronaphthalenes, acetophenones—smell like musk, a widespread smell associated with sexual recognition.

Hogs give off the scent of androsterone in their saliva. The synthetic is used by farmers to induce mating; it can be bought as a spray. Hogs are also used in Europe to sniff out truffles, underground fungi that are eaten as a delicacy. Hogs can find them because they give off the scent of hogs, androsterone. The advantage for the truffle? Spore dispersal, thus reproduction.

The recent discovery of an odor-carrying protein in rats, a protein that almost certainly exists in humans, as well, has led to more speculation and ideas about the workings of the smell mechanism. The protein is produced at the tip of the nose, binds with odorants, and ferries them up to the olfactory apparatus. It may explain why we can smell bell peppers at low concentration and may lend insights into odor blindness. Those who lack the ability to smell certain odors may lack certain necessary transport molecules.

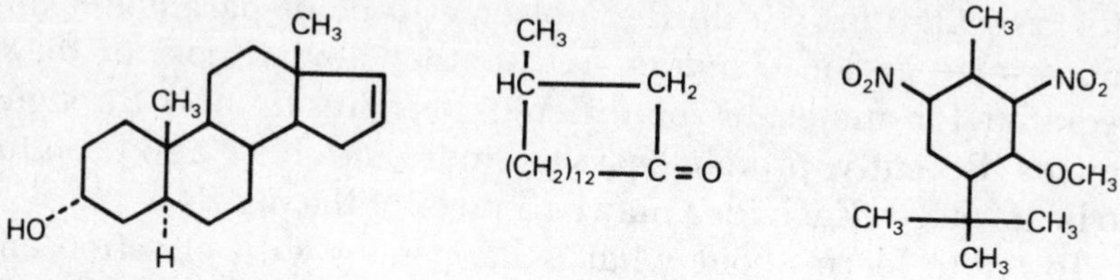

Δ^{16}-Androsten-3-α-ol

Muscone

Musk Ambrette

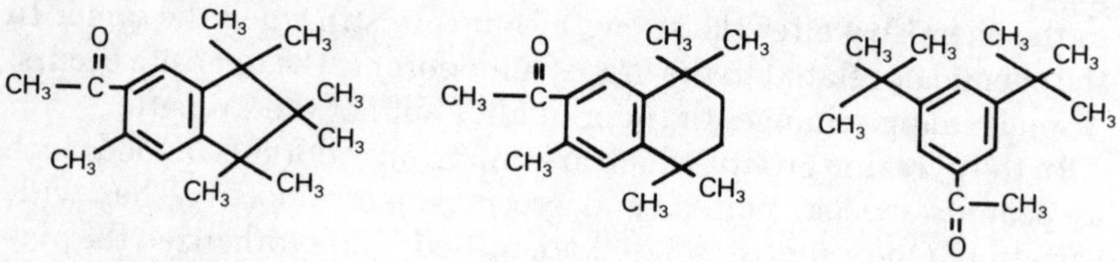

6-Acetyl-1,1,2,2,3,3,5-heptamethylindane

7-Acetyl-1,1,4,4,6-pentamethyl tetrahydronaphthalene

3,5-Di-tert-butyl-acetophenone

A wide range of chemical compounds exhibit the odor of musk, an animal odor. However, an expert perfumer can distinguish subtle differences among them. (From Gustation and Olfaction, *edited by G. Ohloff and A. F. Thomas, © 1971. Redrawn with permission from Academic Press and Dr. Roy Teranishi.)*

Pheromones: Within-Species Coercion

In graduate school, my friend Dave studied cactus mice. One of the observations he made was inhibition of reproduction when female mice were housed under certain conditions with other mice. Although the female mice were reproductively mature and were seen mating, they didn't get pregnant.

Why? On the surface, reproductive inhibition seems maladaptive. But viewed from an evolutionary perspective, it can be a strategy to avoid wasting energy on reproduction when conditions are not ideal. The medium for communicating about conditions—the presence of certain other animals in this case—is chemical.

We began to understand something about the subtlety and extent of chemical within-species communication in the mammals just a few decades ago. The Bruce effect, which has become a classic example, was described by Hilda Bruce in 1959. She discovered that if a mouse that has just mated is exposed to the smell of a strange mouse within 24 hours, the smell of the strange mouse prevents the just-mated mouse from becoming pregnant. To the female, the smell of the strange mouse signals danger and the possibility that she will not be successful in raising her litter. A strange male may kill another male's pups, then impregnate the female himself—adaptive for him and his genes. Ultimately it is adaptive for the female mouse that detects a strange male to postpone pregnancy until the environment is stable. She maximizes her long-term success in this way. The strange mouse smell primes the female's physiological functions that prevent pregnancy.

Chemicals that are secreted externally and influence other animals of the same species are called *pheromones* (in contrast to chemicals such as the skunk smell, broadcast to influence animals of other species—these are called *allomones*). Pheromonal influence may take one of two forms. Primer pheromones, such as the smell of the strange mouse, trigger a chain of physiological events over time. Releaser pheromones trigger an immediate reaction in another animal. Both help to regulate an animal's external environment. Pheromonal communication is widespread in mammals. It is also widespread in the insects.

When honeybee workers sting they release an alarm pheromone, isopental acetate, from their sting chambers. The pheromone attracts other bees and causes them to attack objects in the vicinity of the hive. When more bees sting, more alarm pheromone is produced, stimulating more bee stings. Beekeepers know that one bee sting will often stimulate another bee to sting on the same spot. The notorious Africanized or killer honeybees produce an extremely potent alarm pheromone, one of the reasons for their aggressiveness.

The social insects produce a variety of pheromones that contribute to their organization and control over their environment. Ants produce trail pheromones that they follow back and forth from a rich seed source, for instance, or from the Frosted Flakes. They also produce death pheromones that help keep things tidy, resulting in a kind of grooming of the "superorganism" of the ant colony: Within a few days after an ant dies, the chemical decomposition products of the dead ant act as pheromones, stimulating worker ants to carry the corpse outside the nest. When living ants are daubed with the death pheromone, they are carried to the refuse pile as well. They return to the nest, only to be carried out again and again, until the death pheromone has worn off. An opposite kind of experiment can be done with termites. A termite that has been washed of its identifying chemicals will be treated as an intruder by the soldiers of its colony.

Among the most powerful of the pheromones are chemicals that control sexual behavior. The best known may be bombykol, a chemical produced by the female silkworm moth, *Bombyx mori*, when it is ready to mate. The male moths detect the airborne chemical with their large bushy antennae. The antennae are covered with thousands of sensory hairs containing pores through which the bombykol molecules enter. One molecule of bombykol is sufficient to stimulate a sensory response in the male, that is, to result in neuronal firing. Detecting bombykol, the male begins to fly upwind, following the concentration gradient. The bombykol can attract a male moth from more than three kilometers away. Although a single female contains only about 0.01 millionth gram of bombykol, this is enough potentially to attract one billion males.

Sexual pheromones have been discovered in many moths, and collectors have used them to make their work easier. A female great peacock moth emerged from her cocoon one night in a laboratory, and that same night 40 males came courting, although the species was rare in the area.

Understanding pheromonal communication in insects is contributing to the development of biological control strategies as alternatives to pesticides and their attendant problems. Pheromones bring finesse to the problem, providing a foil rather than a bludgeon. At the U.S. Department of Agriculture Cotton Research Laboratory in Phoenix, where I worked one summer, scientists have been engaged in understanding and developing control methods for "pinkies," pink bollworm moths whose larvae devour cotton bolls. One stratagem uses sticky traps, the stick'em impregnated with pheromones of female pink bollworms. Pink bollworm males follow the scent of the female upwind, only to find themselves stuck in a trap. I remember a South American scientist examining a pheromone trap, a red paper triangle about six inches long, as its use was explained to him. A smile crept over his face. "¡Qué barbaridad!" he said.

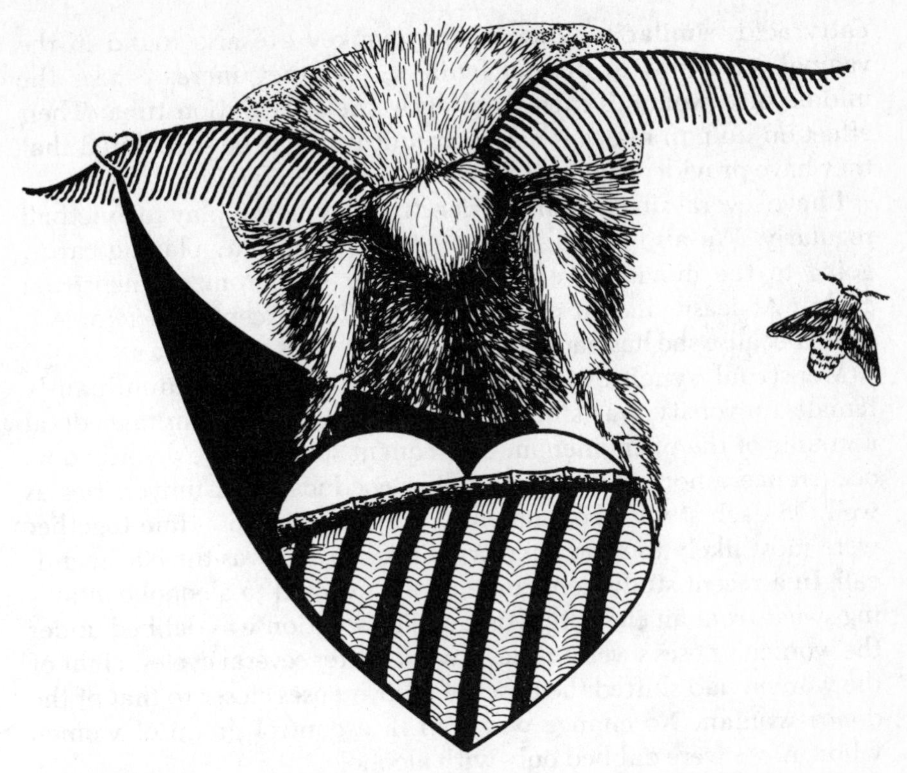

The elaborate antennae of Bombyx mori, *the male silkworm moth, are designed to receive pheromonal signals from female moths who are ready to mate. The female calls chemically, and the male flies upwind, following the concentration gradient of her potent perfume, bombykol. (This sexual dimorphism is also found in mosquitoes—male mosquitoes have fuzzy antennae for detecting romantically inclined females.)*

Sex pheromones in both insects and mammals may have a dual function—attracting a mate and triggering the sequence of stereotyped behavior necessary for mating, another adaptive trait.

The sexual behavior of the male rhesus monkey undergoes rhythmic variations in relation to the menstrual cycle of the female with which he is paired. This sexual behavior may be controlled by the pheromonal action of volatile fatty acids in the vaginal secretions of females. The secretions will stimulate sexual behavior in normal males but have no effect when the male is fitted with nose plugs.

Fatty acids similar to those in the monkey are also found in the vaginal secretions of human females, and they increase near the middle of a woman's menstrual cycle, near ovulation time. Their effect on human males is not clear, but it might be speculated that they have provided subtle clues about optimum fertility.

I have several single women friends with whom I play racquetball regularly. We also spend time together eating out, playing cards, going to the movies. We joke about our synchronized menstrual cycles. At least, all but one of us does. The exceptional female is smug because she has had a hysterectomy.

Menstrual synchrony in women attending a predominantly female university was studied in the 1970s after many anecdotal accounts of the phenomenon. Subsequent studies have described its occurrence among women attending coeducational universities as well. In each study, the women who spent the most time together were most likely to show menstrual synchrony. Was the cue chemical? In a recent study, ten women were exposed to alcohol containing sweat from an eleventh woman. The solution was dabbed under the women's noses several times a week. After several cycles, eight of the women had shifted the onset of their menses closer to that of the donor woman. No change was seen in a control group of women whose noses were dabbed only with alcohol.

Pheromonal menstrual regulation has another facet. Researchers have observed that women who spend more time with men and who have regular sexual relations are more likely to have regular menstrual periods. To determine if a chemical odorant (rather than sexual intercourse) was responsible, researchers collected armpit secretions from males and dabbed them under the noses of women with irregular menstrual periods, those at least 3 days longer or shorter than the normal 29.5 days. The donor males each had the same kind of armpit bacteria, bacteria associated with production of acidic and steroid odors. The women were told only that they were receiving "natural fragrance" extracted in alcohol. The control group, again, was dabbed only with the alcohol. The menstrual cycles of the women who received the male armpit extract shifted toward the norm.

What is the possible advantage of synchronization? Females with regular menstrual cycles seem to have a higher fertility rate than females with irregular cycles, so the substance in male sweat might improve chances of conception. Menstrual synchronization may also be originally a primate adaptation to a hostile environment. In a primate group, some have suggested, synchronization would have resulted in clusters of births, with individuals in a group of same-age offspring perhaps having a greater chance for survival than offspring of variable ages.

Sex Smells

Humans have three regions that are provided with special odorant-producing glands and large odor-diffusing hair tufts: the breast-armpit, anal, and genital regions. These may all be imprinting points in human development. The odor of the areolar glands may be a baby's first smell impression, signaling nourishment. Are men fascinated by women's breasts because of smell memories? One man told me that women's nipples taste and smell "buttery."

Puberty seems to bring on the capacity to smell certain smells. Apparently women who are producing estrogen can smell male sexual smells that are lost on immature girls. Some evidence suggests that as boys become sexually mature they become newly aware of sexual odors of adults and find distasteful the odor of the same-sex parent.

The odor of apocrine sweat appears to be released by the action of bacteria upon the secretion. The odor changes with sexual excitement—the "sex smell" seems to be a matter of common experience. But humans generally nuzzle and kiss necks and ears rather than armpits. I think some of the sex odor must come from saliva and scalp secretions, both male and female.

My friend Jo and I engaged in a pubertal rite when we were both about 12 years old. We were at her house. We locked the bathroom door, lathered our armpits, and used her father's razor to shave the few coarse hairs sprouting from our underarms. That was the beginning. I've been doing the same thing each week for the past 30 years with my own razor, removing the coarse hairs that might provide increased surface area for collection of apocrine secretions and the bacteria that work on them. I admit somewhat sheepishly that I was not among the U.S. females who in the 1960s revolted against symbols of repression such as shaving body hair. Many of my friends, however, grew out their armpit and leg hairs and tossed away their bras, working on liberating themselves from cultural practices that altered their natural selves.

I also have female friends who are repulsed by the horror of feminine "intimate" deodorant sprays, thinking them the ultimate in nutty ideas, not to mention potentially irritating to delicate feminine membranes. But the idea of altering the vaginal odor is not new. Hippocrates in the fifth century B.C. recommended inserting myrrh and aromatics into the vagina to increase the sexual excitement of females and males. Can it be that rather than representing suppression of our sexuality, the new or different odor may be an aphrodisiac, a novel stimulus, erotica to stimulate us sexually?

The role of smells in human sexuality is complicated. One curious human practice is that of removal of the male foreskin, or prepuce, the purpose of which may be as much to hold odor as anything else. Glands of the prepuce produce a cheesy secretion, the smegma, that collects around the foreskin, and the glans penis provides a kind of fixative for more ethereal odorants, much as the fixatives used in perfumes. An analogous secretion collects between the female clitoris and labia minora.

Scratching and Sniffing

"Scratch-and-sniff" is a phenomenon of our modern age. The original technology came from 3M and NCR in the 1950s. Volatile odorants can be surrounded by long polymer chains that form microscopic capsules, which are then stuck to paper. Scratching or abrading the surface breaks the capsules and releases the odorant molecules into the air, where they can be sniffed up the nose. Scratch-and-sniff has been used in perfume ads, cards, children's toys and books, popular collectible stickers, a school nutrition program, a no-calorie picture book of rich foods, even a movie. The smells that can be encapsulated seem infinite: strawberries, lemon pie, root beer, waffles, hamburgers.

Researchers at the University of Pennsylvania have developed a 50-odorant standard test based on scratch-and-sniff that promises to aid medicine and research by lending structure and homogeneity to the diagnosis of smell disorders and studies of smelling abilities.

In 1986, *National Geographic* magazine published "The Intimate Sense of Smell" and included with it a scratch-and-sniff survey prepared in cooperation with scientists from the Monell Chemical Senses Center in Philadelphia, a research institute where 50 or so scientists work on a wide spectrum of questions about taste and smell. *National Geographic*'s circulation, at 10.5 million, is the third largest of U.S. magazines; 1.5 million readers responded, providing an unparalleled trove of data on smell from a broad sample of the population. (And proving that people like to scratch-and-sniff.) Overseas readers also responded.

The six microencapsulated samples that I and others scratched and sniffed in *National Geographic* were androstenone (found in sweat), isoamyl acetate (banana), galaxolide (musk), eugenol (cloves), mercaptan (a sulfur compound added to natural gas—the stuff vultures like), and rose. We were asked to rate the odor's intensity and its pleasant-unpleasantness on scales from 1 to 5. Respondents were given a choice of words (no odor, floral, musky, urine, foul, ink, spicy, woody, fruity, burned, sweet, other) to describe the odors and were asked if they would eat it or apply it to their bodies and if the odor evoked a vivid memory. I found the substances reminded me of new shoes, and especially of a leather handbag I bought years ago in Nogales, Mexico; tough little marshmallow banana candies that I ate as a child; Tabu perfume, which my mother used to wear; spice cookies; new rubber tires; and roses.

Also included were questions about the respondents' handedness (left or right), work environment, use of cologne or perfume, loss of sense of smell, allergies, smoking, pregnancy, and physical illness. Respondents gave their sex, race, nationality, and zip code. Many also sent notes and anecdotes along with their completed questionnaires. The data, entered on computer magnetic tapes, is being made available to researchers everywhere.

In 1987 the magazine published a partial analysis of results. They selected at random 26,200 replies from the United States and compiled the results from 100,000 responses from abroad for an international comparison.

Some of the results: Women not only think they can smell more acutely than men, they can; reactions to odors vary widely around the world; and pregnant women, usually thought to be smell-sensitive, may actually have a diminished sense of smell. Two persons in three have suffered a temporary loss of smell, primarily attributable to colds, flu, and allergies but also from chemicals, pregnancy, and head injury. And 1.2 percent of the population cannot smell at all.

Androstenone, the sweat smell, posed greater problems of identification than any other scent. Seventy percent of women could smell it, compared with 63 percent of men, but neither sex was very successful at identifying it—26 percent women, 24 percent men. Almost all the women and men could smell the banana odorant; half could identify it. Galaxolide, created for perfumers as a synthetic substitute for the odor of musk, formerly collected from the Asian musk deer, showed about the same pattern as androstenone. In the 13 percent of the readers who were unable to smell two odors, androstenone (sweat) and galaxolide (musk) were paired. Androstenone and galaxolide were different smells for me; for Ron, who works in the office next to mine, they smelled identical. ("They're the same," he said. "No, they're not," I said.) Most people could smell eugenol and identify it as cloves; most could smell rose, and the majority could identify it.

Factory workers scored above average in identifying the odor samples and rated the odors as only slightly less intense than average. People working out-of-doors did slightly less well on odor identification but showed higher odor-intensity ratings. Both sexes reached a peak of performance at about age 20 and drifted slowly downhill thereafter. At age 80 women's loss of smell leveled out, but the smell abilities of men slipped even faster. Perception of every odor but one was affected by smoking. In general, smokers found unpleasant odors to be less unpleasant and pleasant odors to be less pleasant.

The *National Geographic* survey disclosed that the stronger the odor, the more likely it was to bring to mind a vivid memory. And just as women found all odors stronger, they reported more memories than men for every odor but gas. Both extremely unpleasant and pleasant odors were as likely to evoke memories, and odor-evoked memories declined with age.

Omni magazine also conducted a smell survey of its readers in

1986 using scratch-and-sniff with five odors and a much different set of questions. More than 20,000 readers responded. Eighty-two percent of readers said they smell their own bodies regularly, and two-thirds bathe seven times a week or more. Favorite smells *Omni* readers reported are roses, lilacs, cedar, pine, rainfall, baked bread, and spring air.

Smell Memories

I once worked for a scientist who studies cactophilic *Drosophila*—wild fruit flies that lay their eggs on decomposing cacti in the Sonoran Desert. Different species of fruit flies dependably choose different species of cacti. We went to the field to find and collect the flies, and the scientist was attracted to the different volatiles from the rotting cacti just like the fruit flies are. Fly-collecting aspirators in hand, we were led through the desert by his nose. I can still see him sniffing and smiling, sniffing and smiling—"No better smells," he said, or something like that. Later in the laboratory I made vats of banana food for the captive flies: overripe bananas, malt, yeast. Today I don't much like overripe bananas, and I don't like banana bread at all.

But people find smells to love in their work. Teachers love schools, journalists love newsrooms, a dairyman loves a nice corral. Others love print shops, hospitals, sawmills, libraries, chemical laboratories, dress shops, taxidermy shops, greenhouses, restaurants, cattle feed yards, laundries, pet shops. Just as in the Gary Larson cartoon of the cows with the air freshener labeled "manure," these places smell right and good to certain people who have fond memories associated with them.

When I was a child, I spent a lot of time drawing and writing, and I still love the smell of paper, crayons, ink, erasers, and pencil shavings. (For some reason, I disliked the smell of modeling clay, and I think it significant that I am not a sculptor.) My father used to sharpen my pencils with his pocketknife. Now that I have an electric pencil sharpener, I sharpen my pencils much more often than necessary, and the smell reminds me of happy hours drawing and writing.

The old rural school I attended smelled of books, paper, crayons, ink, and pencil shavings (as a preschooler, I went once as a guest of an older friend—a first-grader—and I could hardly wait until I was old enough to go every day). Each room had its coal-burning stove. The desks and banisters and swings were slick, coated with oils of generations of students before me. I spent five years there. In fifth grade we moved to a sterile new school. It had no smells. It was not comfortable. Today I live in a house almost a century old with a Franklin stove. I like antiques and most antique stores. They smell good.

If we stay in a particular place for a time, smells give us a sense of belonging to a place in the environment. I spent 18 years in the fragrance of a Utah pinyon-juniper-sagebrush habitat with lilac trees around our house. I missed those smells during my 16 years in Arizona, but I came to love the Sonoran Desert smells, especially the leathery smell of creosote bushes after a blessed rain.

But I most vividly remember the seasonal smells of my childhood. Presented, I associate them with some activity, and they will call up colors and textures and sounds, rather than vice versa. Although I have central heating now, a lump of coal in my stove reminds me of coal burning in the Heaterola in the winter, and my striped pillow and pajamas scorching as they warmed on the top of the stove; the smell of digging in spring flowerbeds reminds me of the wet earth that we drew hopscotch games on in the spring; garden smells remind me of shelling green peas and shucking corn into a zinc tub; the smell of a jar of apricot jam reminds me of a kettle bubbling on my grandmother's wood-burning stove in the fall. I associate the smell of fallen leaves from cottonwood trees with Halloween, because we ate our Halloween candy lying in a pile of raked-up leaves, and I associate the leaf smell with a memory of how the lining of my mouth was raw from consuming the Halloween bonanza of all-day suckers.

Like *Omni* readers, I loved the yeast-working smells of bread-baking day. My mother was a seamstress, and she ironed as she sewed. Different fabrics had different smells. I remember the peculiar ironing smell of a green shirt she made for me, and I am reminded of that shirt—down to the tiny pattern of the fabric—once every few years when I happen to buy an item of clothing that smells the same way when I iron it. It must be an uncommon dye, I think.

In my mind are more than 40 years of food smells and house smells and book smells and people smells, and I never know when one is going to jump out and grab me in the present and take me back. Smells are time machines for all of us.

I began an essay about my grandmother this way: "Infrequently, I have passed closely by a bent old woman at a department store or a concert; wafting behind the old woman, slowly, because she moves slowly, is the smell of Lady Esther powder and the smell of Palmolive soap and the odor of woolens perfumed with sweet aged body oils, given off sparingly over time, and I am flooded with loneliness for my grandmother Rachel, who has been dead some 30 years. . . ."

A man with Listerine breath reminds me of dancing with my first boyfriend (who, significantly, took good care of his teeth and grew up to be a dentist—the signs of our occupations are there if we can only read them). Smell memories of the opposite sex are also as individual as individuals, and I'm convinced that dancing is a socially acceptable way to get close enough to sniff each other out when we're metabolically a little hyperactive, our true selves.

My friend Patty likes to dance, too, and consciously sniffs out men. "They have to smell right to be attractive," she says. "Some men smell like oysters." Patty doesn't like oysters.

My friend Kit disdains men who smell of Old Spice, because "who can get romantic about a man who smells like my *father*?" Yet the author of an article published in *New Woman* reports dancing with a man she has never met before and feeling safe, secure, relaxed, "at home." The reason: "The man smells like my father. To breathe in at his chin or chest level, even through his clothes, is to travel back instantly in my mind to my early years, when the then-dearest man in the world to me, my father, would hold me close. . . . I am drawn almost irresistibly to a man who, at this point, has only one thing going for him: his smell."

Odor Blindness, Anosmia, Hyposmia

We all know people with noses like bloodhounds—I have a sister who fits this category—but almost everyone can't smell something. Specific odor blindnesses are common, and many have been identified as genetic. McKusick's *Mendelian Inheritance in Man*, a catalog of genetic traits, lists, among other things, inability to smell musk, inability to smell freesia flowers, and inability to smell skunk N-butylmercaptan. I can't imagine not being able to smell a skunk.

About 3 percent of the population can't smell isovaleric acid, a key component of sweat. About 33 percent of the population can't smell cineole, the odor of camphor. About 6 percent cannot sense the fishy odor of trimethylamine, secreted in vaginal fluids. About 90 percent of the population can't smell iodocreosol, a substance produced by a reaction of lemon and iodine. Ten to 20 percent of the population can't smell deadly hydrogen cyanide, used in gas chambers (I don't know how this was tested).

An estimated 10 million Americans have chemosensory disorders. More than 200,000 people visit a physician each year with a chemosensory disorder; smell disorders are more common than taste disorders, although the two often occur together.

Anosmia refers to the inability to detect odors, *hyposmia* to reduced ability. In other disorders, a person may detect a foul odor from a substance that to the normal individual smells pleasant. Louis XI of France (1423–1483), "the terrible king" with a long hooked nose, is said to have thought everything about him stank.

Anosmia has varied causes. Some anosmia is genetic. People with Kallman's syndrome have anosmia and lack of sexual interest as part of the syndrome. Some people develop chemosensory disorders after head injuries or after upper respiratory infections. They may be related to polyps in the nasal cavities, hormonal disturbances, dental problems, laryngectomies, medications, or prolonged exposure to toxins such as insecticides. Many patients who receive radiation therapy for cancers of the head and neck later complain of chemosensory disturbances. Recent studies have shown that patients with Alzheimer's disease exhibit deficits both in odor detection and odor identification, which may help diagnosis. Researchers report that surprisingly few patients are aware of their disorder, despite its appearance early in the disease process. Odor deficits are also seen in Parkinson's disease.

However, the predominant problem is a natural rapid decline in smelling ability that typically occurs after age 60. One large study revealed that 60 percent of people 65 to 80 years old had major olfactory impairment, and a quarter of them were anosmic. More than 80 percent of people over age 80 had major impairment, and 50 percent were anosmic.

A person with faulty or reduced chemosensation is deprived of an early warning system that most of us take for granted. Smell alerts us to spoiled food, to fires, poisonous fumes, and leaking gas. Chemosensory malfunctions can affect diet and lead to digestive problems and can wreak havoc with people such as chefs and firemen, whose livelihoods depend heavily on chemosensation.

Several years ago, Mac, a friend of mine from high school, was injured in an accident with a gravel crusher. He and another heavy equipment operator had just made a repair. They turned the crusher on again, and a thick eight-inch pin flew out of the crusher, embedding itself between Mac's eyes, shattering the parchment-thin bones of his sinuses and shearing his olfactory tissue. He was still conscious. He pulled out the pin and spit bone fragments and blood during the long trip to the hospital. His nose has been reconstructed by a plastic surgeon, and he looks the same except for the permanent dilation of one blind eye. But his sensory world has changed. He can't smell, and the result is confusion, he says, although he smiles about not noticing children's soiled diapers. He detects a few things, perhaps through his trigeminal nerve endings, but the sensations they evoke are distorted and not pleasant. He asks his wife, Jo, to tell him what's cooking, then tries to remember that sensation for future reference. Cooking meat smells "awful," and perfume is nauseating, he says, so Jo has stopped wearing perfume.

My friend Jeanne tells me that her grandfather lost his sense of smell when he fell from a swing as a child. As an adult, he worked in a bakery, but because they discovered he couldn't detect the difference between fresh and rotten eggs, he was transferred from the cake-baking section to a section where no eggs were used.

Without the clues provided by smell (and vision), eating an apple and a raw potato are indistinguishable experiences. The texture and taste are the same; the volatiles provide the identification. Volatiles also provide the identification for tea and coffee and chocolate. By floating a drop of vegetable oil on top, to prevent the drinks' volatiles from escaping, or by holding my nose, I can get some idea about Mac's deprivation and the frustration he feels but not of the total picture. People who can't smell don't call up memories with present smells, and they aren't making new smell memories.

Artificial Noses

Chemists use procedures known as gas chromatography, high-pressure liquid chromatography, and mass spectroscopy to separate and identify chemicals in the laboratory. But artificial "noses," instruments that are able to detect a number of different gases, may eventually aid faltering human noses. One, consisting of gas-sensitive semiconductors, is being developed at the Robotics Institute of Carnegie-Mellon University and is predicted to someday make an ideal air-pollution watchdog and to revolutionize mass-production methods for drugs, chemicals, and processed foods. Another has been developed by Sandia National Laboratories to detect a range of gases that indicate corroding metal or leaks in stockpiles of nuclear weapons, but its inventors predict its usefulness in other situations as well. It is about the size of a pea and consists of six tiny diodes etched in a silicon wafer, each with a different combination of metals. Various gases that contain hydrogen lower the resistivity of the metals, causing a flow of current that can be measured and used to set off an alarm. It is sensitive to several gases at concentrations of only a few parts per million. Future versions may be worn like a wristwatch, to alert anosmic people to gas leaks, fires, or spoiled meat.

Environmental Odors

My friend Diane spent three months trekking about in Nepal. She took only occasional sponge baths during that time, and like another friend who spent a winter trapping in Alaska without bathing, she says her body oils reached a state of equilibrium. Her skin was smooth and supple, her hair sleek. Her time in Asia was a rich olfactory experience. She was overwhelmed with the different scents of people, food, sewers, flowers. "When I came home and got off the plane in Seattle," she says, "it was strange. There were no smells."

Although some smells are certainly the same around the world, different cultural and environmental practices, different plants, and different foods result in different environmental smells. In the

United States, we make a willful effort to manage our olfactory environment. Advertisements on television and in magazines promote the idea that indoor and outdoor air should generally be odorless. Odors produced by warning agents, perfumes, and foods are welcome only at certain times and in certain places.

Generally, a lack of odors means better public health practices—control over sewage and garbage disposal or air pollution—but there are other good reasons for seeking control over the olfactory environment. The presence of extraneous olfactory stimulation, which may or may not be objectionable in itself, becomes objectionable when it interferes with receiving desirable chemical stimuli.

People Smells

Alexander the Great is said to have had wonderful breath and to have exuded a fine body odor that permeated his clothing. Walt Whitman, no humble man, proclaimed that his own (unadorned) armpits smelled "finer than a prayer."

Helen Keller, who was deaf and blind, had an extremely fine-tuned nose. She remarked about the "dear, unmistakable" odors of those she loved. She said that her nose helped her to learn much about people. She could deduce the work they engaged in from such things as wood, iron, paint, and drugs clinging to their clothing. "When a person passes quickly from one place to another," she said, "I get a scent impression of where he has been—the kitchen, the garden, or the sickroom."

In old country houses Keller could smell layers of odors left by a succession of families, plants, perfumes, and draperies. She said that people had "personality" smells. Infants lacked a personality scent, and adults who lacked a distinctive person scent she seldom found lively or entertaining. Keller found male scents stronger, more vivid, and more differentiated than those of women.

Recent studies have shown that mothers recognize the scent of their own babies and can pick out clothing their children have worn. Babies will also prefer underarm pads worn by their mothers over those worn by other women (or their fathers), indicating smell recognition.

Because people are all a little different chemically and metabolically, we should be expected to have different personal odors. Study indicates that husbands and wives do not smell alike, so diet does not homogenize a couple's smell. Perfumers are reputed to be able to smell differences between skin and hair colors. Democritus writes of being able to distinguish virgins and nonvirgins by their odors. Medieval authorities claimed to be able to discriminate the odor of chastity in both sexes (resembling the chemical ionone, used in perfumery and flavoring) and that of unchastity (resembling boiling starch). In 1886, Augustin Galopin, in *Le Parfum de la Femme*, classified women according to smell. Blondes, he said, smelled of amber; women with chestnut brown hair of violets; brunettes of ebony wood; and redheads smelled "peculiar." (Shades of the bedbug story.)

Some psychiatrists have long claimed to be able to smell schizophrenics; the substance involved has been identified as trans-3-methylhexanoic acid in their sweat. On the other hand, some schizophrenics, whose senses seem to be "turned up," claim to be able to smell hostility.

Physicians associate certain odors with certain illnesses, especially diseases resulting from inborn errors of metabolism, in which the usual metabolic pathways are subverted because enzymes are faulty or lacking. The urine of children with *phenylketonuria* has a sweet almond smell. In phenylketonuria, the enzyme phenylalanine hydroxylase is missing and phenylalanine builds up, which is toxic to the brain in the early years of life. Newborns are now routinely tested for the malady and treated by a diet low in phenylalanine. Not many years ago, however, the disease went untreated, and the resulting retarded children were often institutionalized, imparting to institutions their sweet almond smell.

Maple syrup urine disease is another metabolic disorder. And diabetics, who are metabolizing fats into ketones because they lack the hormone insulin to move sugar into their cells where it can be used, have a characteristic fruity ketone odor to their breath (an odor in the family with acetone, nail polish remover).

Francis Bacon and others wrote of the plague's smell of "mellow apples." Other smell descriptions of disease: typhus, mouselike; measles, like freshly plucked feathers; scarlatina, like fresh hot bread; eczema and impetigo, moldy; and nephritis, like chaff.

Healthy people should be free of such strange odors, and this perhaps provides a subliminal criterion for humans in the search for potential mates. Odor estrangement from one's group is not good. A few years ago, my friend Barb, a hospital dietitian, worked out a special diet for a patient with "fish odor syndrome," or trimethylaminuria. The patient's problem was related to excessive dietary choline, resulting in a buildup of trimethyl alanine generated by gut bacteria that could not be metabolized by the liver, again because of a lacking enzyme. His faulty protein metabolism resulted in a strong body odor of rotting fish, unpleasant to those around him. The diet improved his odor.

Why Do Humans Wash?

As I write this, Cruiser the cat licks himself and combs his fur with his teeth. (When he begins to sharpen his claws on my fake Persian rug, I'll interrupt him.) The birds splash in their drinking water and preen their feathers, zipping them through their bills. Soon I will take my daily shower and file my nails.

We animals spend a lot of time grooming.

For humans, grooming generally means washing of skin and of clothes. My encyclopedia tells me that bathing, "serving both for cleanliness and for pleasure, has been almost instinctively practised by nearly every people." Ancient Egyptians, Greeks, Persians, Jews, Turks, Orientals, and American Indians bathed. The Romans, especially, built elaborate and ingenious baths in which they steamed themselves, scraped their skin with curved scrapers known as strigils, and anointed themselves with fragrant oils.

If chemical communication is so basic, why do humans persist in washing, scrubbing, and deodorizing their bodies? The answer seems to be the same as the answer for grooming behavior in cats and birds: to get rid of ectoparasites and the secretions and shed cells that provide potential substrates for parasites or bacteria. For humans, washing or airing clothes—subjecting them to the deodorant action of ultraviolet light—is an important adjunct to bathing to break down or get rid of built-up organic molecules and to avoid reintroduction of parasites.

Although bathing has sometimes been given up (under the guise of saintly behavior in the Middle Ages) or has been discouraged as debilitating (my grandmother felt that frequent bathing had a "weakening" effect), it is a ubiquitous part of the human behavioral repertoire. Bathing has varied over the centuries with cultural practices, religious belief, location of hot springs, technology for transporting and heating water, climate, and central heating. In cold climates or cold winters when hot water was not just a turn of a tap away, human bathing was more difficult. My parents talk of the trouble involved in heating water for bathing, especially on wood-burning stoves during cold winters. When people possessed fewer changes of clothes, and washing clothes was more difficult and less frequent, we can assume that human odors were much more apparent. But these were not just pure chemical communication odors; they must have included unpleasant odors associated with the growth of certain bacteria and parasites and with rashes or sores, signals of unhealthiness.

Bathing, along with the sanitary disposal of human feces, has unquestionably contributed to better human health and less itching and scratching. Parasites quickly gain when bathing and clothes-washing are suspended, especially when people are crowded together. Without bathing (and even sometimes with bathing, as elementary teachers who have dealt with classroom lice epidemics will attest), head and body lice, those little medieval "pearls of poverty," have a heyday. History, art, and literature abound with examples of human life with ectoparasites and attempts to foil them. Egyptian priests and others have shaved their heads and bodies to keep the lice under control. In the twelfth century, Thomas Becket's body was observed to "boil over" with lice as it was being prepared for burial, the lice bailing out as his body cooled. My friend Kris tells me the pleated collars, the picadills, in paintings by the Dutch masters were designed as much as head lice-catchers as for decoration.

Perfumery

I have often heard remarks that humans are peculiar because they wash off their own body odorants and replace them with the odorants of other animals or plants. We like to titillate our noses with pleasant odors and to use odors for creating moods and atmosphere. The practice of perfumery (from the Latin *per fumus*, "through the smoke") may date back to 6000 B.C., associated with religion, animal sacrifice, and pleasing the deities (to solidify ritual and make smell memories?). The temple of Baal in Babylon burned two-and-a-half tons of frankincense a year.

Perfumes were used everywhere to mask unpleasant odors and for adornment. Shakespeare, describing Cleopatra on the Cydnus River, emphasizes the clouds of fragrance that encircled her. The ancient Romans had festivals of flowers. The Romans also had guilds of perfumers. Nero burned a year's production of incense at the funeral of his wife Poppaea. Perfumery at one time became so excessive, and associated with debauchery, that in the second century the Church condemned personal perfumery among Christians.

We do love flower odors (produced as attractants for pollinators, generally insects), but some of the artificial scents that humans use to replace the natural odors they remove by washing may not actually be so artificial. For a long time humans have appreciated the animal sexual attractants musk, from the Asian musk deer (hunters are said to note its sweet smell in the forest), and civet, from the African civet cat. These substances are widely used in perfumes. Civettone and the steroid ketone androstenone are similar in structure and smell. Musk smells are like those of the human sex pheromones. Ambergris from the stomach of the sperm whale and castor from beaver glands are also used in perfumes. The animal odors are extremely potent and are used in minuscule amounts as fixatives, acting as heavy anchors for lighter molecules in mixtures.

In an attempt to articulate the inarticulate sense, people who produce fragrances talk of fragrance "notes" and "chords," of using natural and synthetic fragrances artistically to evoke flowers, fruits, spices, precious woods, animals. Blended with musk, amber, mosses, they transform into a "symphony" of harmony. In an article about fragrance description, perfumes are said to emphasize a woman's charm, her elegance, or natural freshness; or they become scents that stress a man's virility, sportsmanship, or feelings. The article goes on about ambiance: The rose olfactory note suggests softness, femininity, and sensitivity; spicy notes of clove, coriander, and cinnamon express exoticism, light, warmth, and arrogance; violet notes lend elegance and distinction; lily of the valley is reminiscent of spring, morning dew, youth, lightness, and delicacy; sandalwood confers soft effects, vetiver expresses warmth, patchouli brings voluptuous sensations; balsamic and animal notes can be calming, mysterious, sweet, or sensual; fruity notes evoke sparkle and mellowness but are also voluptuous and glamorous; lemon expresses freshness, sun, beach, sea, wind, waves, the South—add spices to it and get the Caribbean islands.

These are emotional descriptions as much as anything. Scientists are finding that scents can in fact have druglike emotional effects: easing pain, calming anxiety, reducing hunger, ending bouts of insomnia. Aroma therapy is in its infancy as a science, but scientists speculate that odors may be used in the same way as drugs with the special advantage of their direct route to the brain.

The fragrance industry is big business. International Flavors and Fragrances, Incorporated, a company headquartered in New York City with subsidiaries all over the world, is the industry leader. In addition to all the regular things one might expect them to produce, they have been called upon to make the odor of fresh-baked bread for a bakery that was being overwhelmed by a pizza establishment next door, the odor of a cave for an amusement park, and the essence of slum (with garbage and urine) for a Smithsonian exhibit.

A Life of Odors

We each lead a life of odors. From the time we are suckling babies, odors become meaningful through association with experiences, and associations continue to appear and reappear throughout our lives. Odors have a stimulating effect, but research has shown that very young children are less likely to dislike certain odors than are older children. They need experience. Then with puberty and hormones, a different olfactory world sneaks up on them.

That's about when, like Cleopatra and Alexander the Great, we each make our greatest contributions to the fragrance industry. I looked into my bathroom cabinet recently. Even my meager contribution is significant. I have my favorite fragrances to wear, and I found some lotions and potions to throw away because I don't like how they smell. They were mostly given to me by others, tested by different noses. (Add perfumes to the list of Christmas no-no's, I thought, along with puppies and paintings.) I also found a man's cologne that I once bought to dab on my pillow to remind me of someone who was far away. I had forgotten him. The smell contained memories.

Sans teeth, sans eyes, sans taste, sans everything.

—William Shakespeare

The smell and taste of things remain poised a long time, like souls, ready to remind us. . . .

—Marcel Proust

Dewy flap; goalie to the throat, cheek-char; mopgum; a muscle from whose trembly tip poetry drips, like saliva.

—Richard Selzer

The Owl and the Pussycat went to sea
In a beautiful pea-green boat,
They took some honey and plenty of money,
Wrapped up in a five-pound note.

—Edward Lear

You can catch more flies with honey than with vinegar.

—My mother

F·I·V·E

Taste: Accounting for It

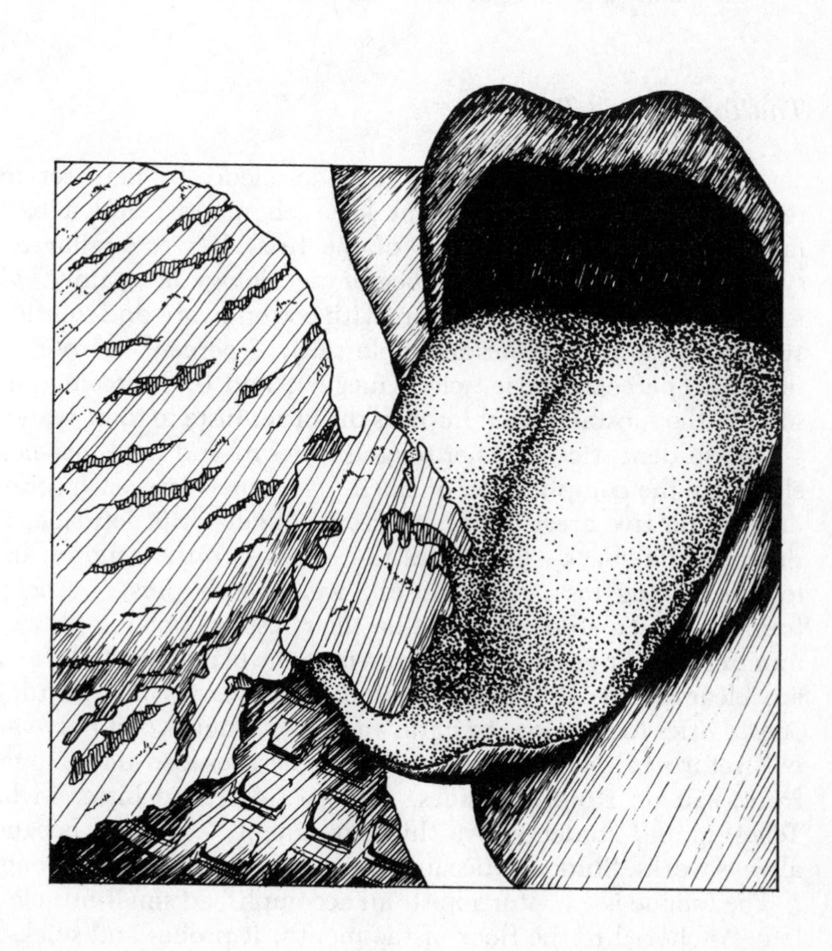

TASTE (from Lat. *taxare*. to touch sharply; *tangere*, to touch), in physiology, the sensation referred to the mouth when certain soluble substances are brought into contact with the mucous membrane of that cavity. By analogy, the word *taste* is used also of aesthetic appreciation and a sense of beauty, commonly with the qualifications "good taste" and "bad taste."

The Tongue and Taste

Several years ago I subscribed to a do-science-by-mail service. I received a different experiment kit each month, and a tongue-mapping experiment was one of the first. It was similar to the typical experiment done in a beginning biology laboratory, which requires students to work in pairs with cotton swabs and solutions of 10 percent sodium chloride (table salt), 5 percent sucrose (table sugar), 1 percent acetic acid (vinegar), and 0.1 percent quinine sulfate (the substance that imparts the bitter taste to tonic water).

One student sticks out her tongue, dries it, and holds her nose to eliminate the complication of smell. The other dabs each solution, in turn, on five areas of the stuck-out tongue—tip, left side, right side, middle, back—and notes the response on a diagram of the tongue: strong taste (two pluses), weak taste (plus), or no taste (minus). When students combine their results, putting pluses and minuses on four giant tongues drawn on the blackboard, differences are clear in the distributions of the four tastes on the tongue. Sensitivities to solution strengths vary, and tongues certainly vary (I will return to this), but there it is in the averaged data: sugar in front, salt on front and sides, sour on sides, and bitter in back. Teachers and students love this experiment. Teachers, because it always works. Students, because they have discovered their tongues.

The tongue is a contortionist, an accomplished small-muscle athlete. Anchored to the floor of the mouth, it probes and snicks and sweeps about. It narrows and sticks out, it fattens and retracts, it

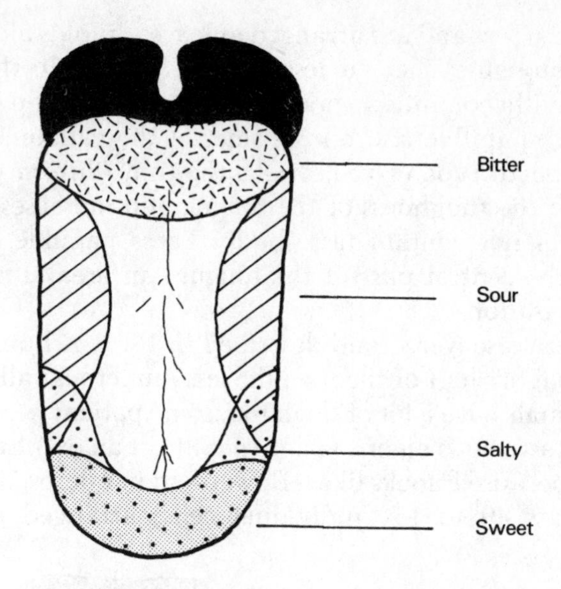

Tongue mapping reveals areas sensitive to sweet, sour, salty, and bitter.

curls, it manipulates food, wipes the lips, and licks the fingers. It clicks against the lips and teeth and palate, making the consonants of speech. It participates in kisses of passion.

The tongue provides another barometer too. A "furry" or "coated" tongue belongs to a sick person. A black tongue can indicate colonization by yeast in a mouth ecologically out of balance, perhaps because antibiotics have killed the normal faunal and floral population.

The tongue is covered with a mucous membrane, folded on the upper surface to form little peglike projections known as *papillae* (from the Latin for "nipple"), which in turn contain the taste buds, receptors for the chemical sense of gustation. The taste bud-containing papillae are of three types: fungiform, vallate, and foliate.

Fungiform papillae, shaped like little mushrooms, are scattered over the upper surface of the tongue. With the naked eye, they appear as red spots, because of their rich blood supply. The *vallate* papillae are shaped like flat mounds surrounded by a trench. They are much fewer in number than the other types of papillae—humans have only 7 to 12 of them, restricted to an inverted V-shaped zone across the back of the tongue near its base. The *foliate* papillae are also few and are arranged as leaves or folds along the back edges of the tongue. They are less prominent in adults than in children. Each papilla contains several taste buds, located on the tops of the fungiform papillae and on the sides of the vallate and foliate papillae. A fourth type of projection, the thin *filiform* papillae, contributes to the roughness of the tongue and increases its surface area but does not contain taste buds. These papillae are most numerous on the central part of the tongue, an area almost devoid of the taste sensation.

The taste buds were discovered and described in 1867. A human has about 10,000, roughly half of them on the less numerous vallate papillae. (Other animals have a lot of them too; some patient person once counted 1,760 taste buds on one vallate papilla of an ox.) Each taste bud, so called because it looks like a flower bud in microscopic cross-section, contains 40 to 150 individual cells, arranged like

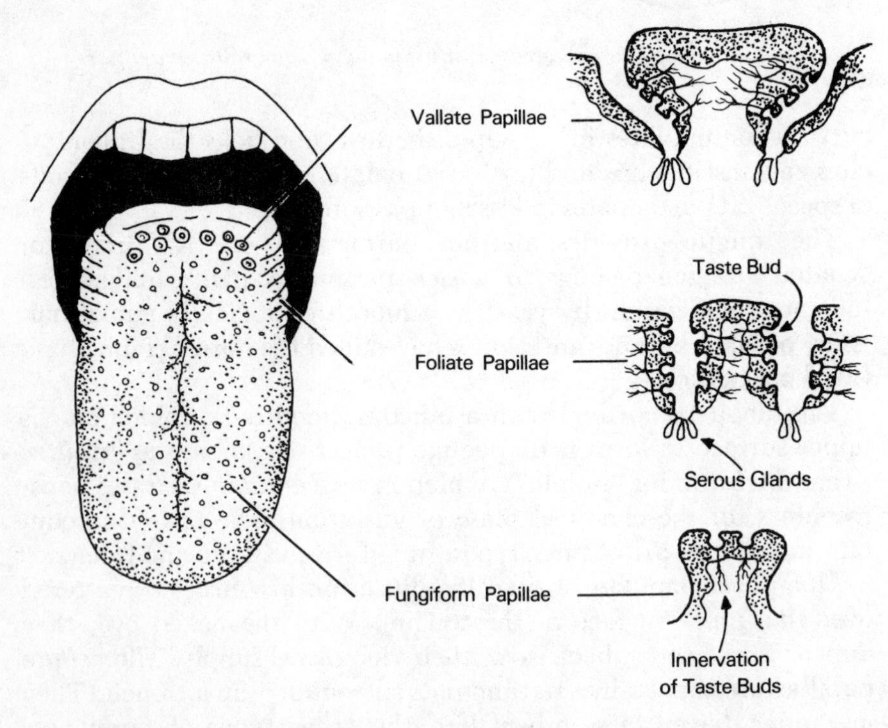

Three kinds of taste-bud-containing papillae are found on the tongue. Chemicals must be in solution to be tasted, a remnant of our evolutionary past.

staves of a barrel. The sensory cells are sensitive to chemical stimulation. A few supporting cells provide props, and the basal cells give rise to new sensory cells. The life span of a sensory cell can range from a few days to a month; the average is ten days.

The taste bud is slightly recessed below the epithelial layer of the papillae. Chemicals, dissolved in water supplied by the salivary glands, enter a pore into a fluid-filled space over the taste bud, where they contact projections of the cell membrane at the tip of the bud. The contact causes a response that is in turn transmitted to the 50 or so nerve fibers entering each taste bud. When new sensory cells are replacing old ones, new connections are formed between new cells and the nerve fibers, a flux that may account for a changing taste profile. Fibers from the taste buds collect into two of the cranial nerves. Touch sensations from the tongue are carried by still another nerve. The mouth is a busy place. It assists the appetite, the body's crying-out for energetic help.

Eating engages all the senses in some way. We see an apple, touch it, smell it, and hear it crunch as we chew. The taste buds respond to apple chemicals, among them malic acid and the sugar fructose. Stimulation of the taste buds is joined by stimulation of other receptors in the mouth—receptors for pressure, heat, and cold. Vaporized chemicals from food enter the nasal passages, both through the nose and via the back of the mouth, and stimulate the olfactory processes. Pain receptors are sometimes stimulated. "Taste," actually "flavor," is a combination of all these sensations. Taste in the narrow sense, however, is categorized into the four basic sensations of the tongue map: sweet, sour, salty, and bitter.

The sweet taste, for instance, is stimulated by a host of simple sugars, among them sucrose, glucose, fructose, maltose, and saccharose. It is also stimulated by some unlikely chemicals, such as the nonnutritive saccharin, the amino acid D-leucine, lead acetate ("sugar of lead"), chloroform, and beryllium chloride. (The first name given to the element beryllium was glucinium, which means "sweet.")

The sour taste is stimulated by such substances as hydrochloric acid, acetic acid, citric acid, malic acid, and tartaric acid. The salty taste is stimulated by such salts as sodium chloride, ammonium chloride, magnesium chloride, and sodium fluoride. And the bitter taste is stimulated by a long list of chemicals, among them

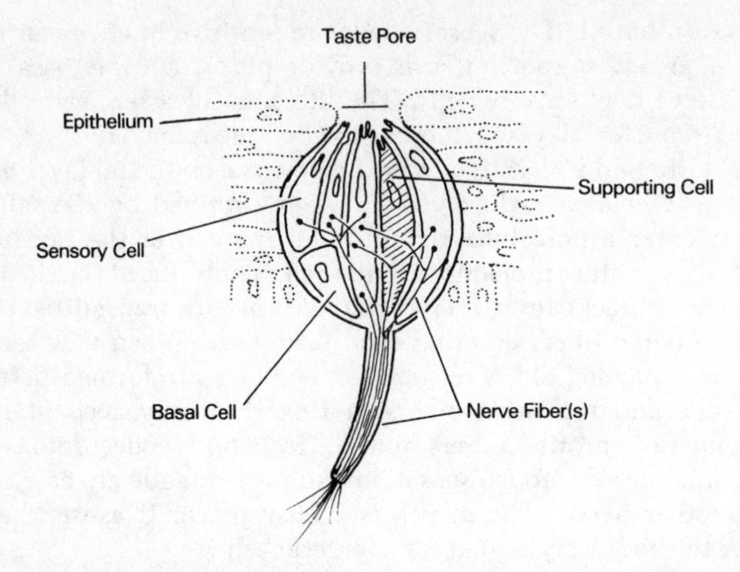

Taste bud cells are constantly renewing themselves from dividing basal cells. The glands at the base of the taste bud secrete water to wash the chemicals away.

quinine sulfate, nicotine, caffeine, L-leucine (the left-handed version or mirror image of the amino acid D-leucine), morphine, and magnesium sulfate. Each taste has its own adaptive significance, as we shall see.

A substance must be dissolved in water to stimulate the taste buds. Stick your tongue out, dry it with a towel, and put some sugar granules on it. Nothing—until the tongue starts leaking water, and the sugar dissolves. The *detection threshold* is the minimum concentration in water at which a substance can be tasted, and the *intensity* is a measure of the strength of the taste sensation. The threshold can be measured carefully, and rating scales are developed from standards to use for intensity measures. Individual tastes and thresholds vary over time, but in general, substances with a bitter taste are detected at very low concentrations. The thresholds for common simple sugars, with molecules containing three carbons or six carbons, are higher than those for synthetic sweeteners. That is, the synthetic sweeteners are many times sweeter than natural sugars and are detected in minuscule amounts. Threshold concentrations for acetic acid (sour) and table salt are about the same as those for the sugars.

A sensory cell in the taste bud is stimulated when a chemical contacts a receptive protein molecule embedded in the cell membrane. This contact, or adsorption, creates a temporary chemical/protein complex and causes a change in permeability of the cell membrane, which results in "firing" of the cell. Some substances will temporarily block certain taste qualities. For instance, gymnemic acid, obtained from leaves of the Indian plant *Gymnema sylvestre*, will block the receptor molecules for sweet substances, rendering them temporarily tasteless, so that table sugar tastes like sand. Could the plant chemical be trying to tell a potential predator that the plant really isn't good to eat? Gymnemic acid will also block bitter taste, somewhat later than the sweet taste.

Taste buds are also found in the soft palate, the *uvula* (the bell-clapper that hangs at the back of the throat and vibrates when we say *aah*), the *epiglottis* (the flap that covers the opening to the windpipe when we swallow), sometimes even the vocal cords. Children have more extralingual taste buds scattered about than do adults. Claude-Nicholas LeCat's *Physical Essay on the Senses*, published in 1750, describes two tongueless children who, however much trouble they may have had manipulating food, nonetheless had the ability to taste. One child was born without a tongue, the other had lost its tongue from gangrene as a result of smallpox.

Extralingual taste buds diminish over time. These young taste buds are presumably an adaptation for a critical time when children are learning about foods and might explain why children are such picky eaters. I remember being picky myself, and just when I had recovered from it, I was cursed with providing meals for a child for whom the green pea torture rivaled the Chinese water torture. The revenge comes finally to the older generation, of course, when the curse is passed on. Many mothers are familiar with children who roll peas around on their plate. One man tells me he did it as a child but could also make them disappear: Like the prisoners digging an escape tunnel and hiding soil to be disposed of later in the exercise yard, he hid his vegetables in the cuffs of his pants, when his warden wasn't looking, to be disposed of later, outside. His system worked for peas, green beans, and carrots, he says, but not so well for creamed vegetables. This man is now a child psychologist who likes peas and relates well to children.

We are not born loving green peas, but research has shown that children are born with a desire for sweet taste; that is, they will preferentially suck a sweet solution. Their first food, milk, contains lactose, milk sugar. A sweet taste generally means a food is calorically and nutritionally good for us. Undoubtedly, it has been adaptive for children to prefer sweet foods or bland foods and to avoid novel foods or foods with a strong, distinctive taste, foods that may not be good for them. Children don't have to be persuaded to munch on cookies and crackers but are no more interested in broccoli than Cruiser is. Their taste buds reinforce this behavior early on.

Over the long human dependence period they learn, abominably slowly, from their experienced omnivore parents about other acceptable foods and add foods to their diets, so the repertoire increases. This pattern of weaned young learning from parents about palatable foods, especially palatable plants, is widespread among the mammals. For instance, range scientists know that the young of grazing animals follow along with their mothers, sampling the plants that the mother eats. (Some of the range plants are poisonous at certain times and when consumed in large amounts, but as a minor part of the diet of a large plant eater, they may pose no problem.) Avoidance of novel foods is known as food *neophobia*. Experience, through which humans and other animals become familiar with a food, plays an important role in shaping food preferences. Two-year-old children exposed to novel cheeses and fruits over a period of time increased their preference for them. A food-choice study with children four and older showed that sweetness was the primary determinant in their food preferences.

Human mothers may discover that if a little applesauce is mixed with the green peas, the baby will eat them more readily. If the gods are smiling, in the not-too-distant future the green peas take on a desirability of their own.

From Greek Tongues to Psychophysics

The Greek physician Alcmaeon (ca. 600 B.C.) believed that the tongue admitted tastants to the "sensorium"—the seat of sensation—through tiny pores. Democritus (460–370 B.C.), credited with originating the theory of atoms, explained taste as resulting from the shapes of things making up what was being tasted. He imagined sweetness as consisting of round particles; sourness as consisting of angular particles that pricked the tongue; and bitterness as round particles covered with little hooks that snagged the tongue.

Aristotle (384–322 B.C.) listed the taste qualities of sweet, sour, salty, and bitter—plus astringent, pungent, and harsh. The four basic qualities have stood the test of time, with some quibbling about additional qualities. The Roman physician Galen (A.D. 180–200) agreed with Aristotle's interpretation; the Muslim physician Avicenna (980–1037) listed the four basic qualities plus insipid; the French physician Jean Fernel (1497–1588) added fatty to the list.

The discovery of animal electricity—credited to Luigi Galvani (1737–1798)—galvanized the scientific study of taste. Galvani had found that an electrical machine would cause muscular contraction in the bodies of recently killed frogs. Alessandro Volta (1745–1827) experimented with the production of taste sensations by touching together a piece of zinc and a piece of copper, which had been found to conduct electricity.

Psychophysics, the combination of physics and psychology, is considered to have been born as a discipline with the work of the German physicist Gustav Fechner (1801–1887). The first laboratory of experimental psychology was established in 1879 in Leipzig, Germany, by Wilhelm Wundt (1832–1920), who had been influenced by Fechner and who was a student of Helmholtz, the physicist who invented the ophthalmoscope. The laboratory must have buzzed with activity. The senses were stimulated electrically and the reported psychological effects recorded. With the studied application of mathematical measurement to sensation, the psychophysicists discovered much about the senses, including taste. They discovered the existence of taste thresholds and described sensory adaptation, the phenomenon of lessened response or the cessation of a response to a particular stimulus after a period of repeated exposure.

Sensitivities to the four tastes (they argued for some time about whether more existed) were shown to be differentially distributed across the tongue, and the qualities were known to disappear in a certain order after the application of topical anesthetics. From 1930–40, one experimentalist, Helmut Hahn, studied 15,000 thresholds for 108 taste substances on 43 subjects and measured the effects of adaptation on taste thresholds. The 1956 list of Yngve Zotterman still includes the four basic tastes of Aristotle—sweet, sour, salty, bitter—plus, in some species, the taste of water. (The Japanese include the taste of monosodium glutamate, which they use as a flavor enhancer.) After Hahn, studies of taste psychophysics became almost exclusively an American discipline, with the American scientists tracing their intellectual lineage to all those tongue-testing Germans.

Sensory evaluation of food has become an important speciality within the field of food technology, and understanding of taste responses has broadened. Electrophysiology has been refined and has been used to measure the responses of individual nerve fibers in many different kinds of animals. The taste nerves are tapped with an electrode, the tongue stimulated, and the neural response recorded. Electrical stimulation and recording of responses, using animal subjects, have established that some nerve fibers respond to acid only, some to acid and salt, and some to acid and quinine. The most specific nerve fibers seem to be those that respond to sugar solutions, and these are particularly characteristic of primates, the fruit-eating group to which we belong. The failure of nerve fibers to fall into distinct categories has led to the idea of a taste profile; that is, one substance may result in firing of the four different kinds of fibers, but a particular taste shows a particular pattern, such as mostly sweet or mostly bitter.

Magnitude estimation studies support the taste profile idea. The subject's tongue is rinsed with water, and a certain solution is flowed over the tongue for five seconds. Subjects estimate the intensity of each taste. In this procedure, some substances have one taste that predominates, but many others result in combinations of the four tastes. The four substances we used in tongue-mapping each come close to having only one taste, but potassium chloride has substantial bitter and salty components. Sodium nitrate results in a taste made up of salty, sour, and bitter.

A *Time* magazine food critic who took a taste test a few years ago wrote something reminiscent of Democritus: "It was interesting to notice how the tastes literally 'felt' as they were being washed over the tongue. Salt and sweet were warm and pleasant; sweet was the most relaxing and salt was exhilarating. Bitterness curls the edges of the tongue. Sour felt icy and caused the surface of the tongue to contract."

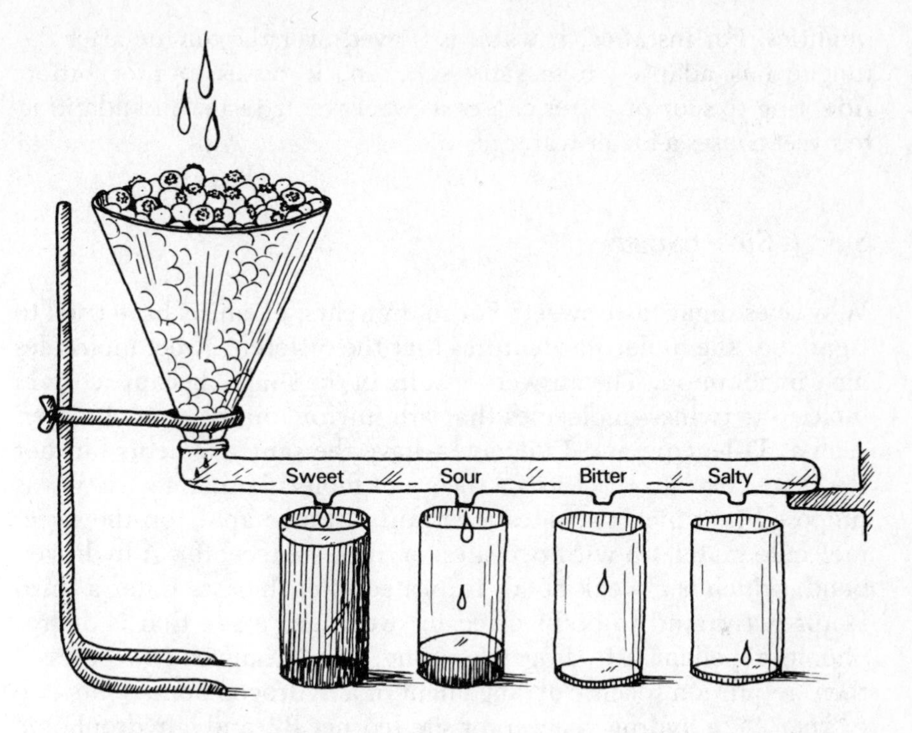

Tastes are not usually pure, but rather contain a combination of the four qualities. Ripe blueberries, for instance, have about nine times as much sugar as acid. As they change to overripe, the sugar-to-acid ratio increases.

The psychophysicists have found that some species (frog, cat, dog, pig, rabbit, squirrel monkey) have a taste response to pure water, while others (rat, hamster, goat, calf, macaque monkey) do not. The reasons for this are not clear; however, saliva components can affect taste and can vary over time, and the water animals find to drink in nature is not distilled water. Also, the taste sensation produced by distilled water seems to be a response to the removal of sodium chloride in saliva. In some people, the saliva contains enough sodium chloride to make water distinctly bitter-sour. Others' saliva contains so little that water is tasteless. But for many, saliva is intermediate. The water is detected, but the taste cannot be recognized. Perhaps this is the source of the "insipid" taste listed by some early taste classifiers.

When the tongue is adapted to certain substances, water can be given a kind of echo taste, taking on the taste of any of the four taste qualities. For instance, if water is flowed over the tongue after the tongue has adapted to a salty solution, it tastes sour or bitter. Adapting to sour or bitter causes a sweet water taste, and adapting to sweet causes a bitter water taste.

Sweet, Sweet Sugar

Why does sugar taste sweet? For many years, chemists have tried to figure out the molecular features that the different sweet molecules have in common. The answer must lie in the shape, because certain molecular twins—molecules that are mirror images of each other, such as D-leucine and L-leucine—have the same elements but not the same taste. In the 1960s a theory of molecular connectivity was proposed in which two sites a certain distance apart on the sweet molecule match up with two sites on the taste receptor. A hydrogen bond, which is a weak bond, is formed in each case. Later a third factor was found to be involved in sweetness: a site that is hydrophobic, repellent to hydrogen bonding. The "triangle of sweetness," then, requires a specific arrangement of a hydrogen bond donor site (corner A), a hydrogen acceptor site (corner B), and a hydrophobic site (corner C). The three sites must be separated from one another at certain distances. Because molecules are three-dimensional and flexible, the triangle is not always easy to locate. In table sugar, for example, eight groups could serve as corner A, 11 atoms could be corner B, and three groups could be corner C.

Because artificial sweetening agents and low-calorie foods have taken on significant commercial and economic interest, the theories of sweetness have given scientists some leads for developing and testing a series of molecules that should taste sweet. It's more complicated than it sounds, however. Novel sweeteners may not hold together well. As their atoms come apart, different molecules with different structures are produced, and detection of bitterness is not far removed from sweetness.

The natural sugars (saccharides) are widespread and varied. They occur in significant amounts in honey, fruits, vegetables, and milk—desirable human foods. Many sugars have been identified, and about a dozen of them are common dietary sugars. By tasting the quality "sweet," we can discern the approximate amount of sugar in food; it also gives us and certain other animals a means by which we tell whether fruits are ripe and whether unknown foods are safe to eat and nutritious.

Simple sugars are small, as organic compounds go, consisting of carbon, hydrogen, and oxygen in a 1:2:1 ratio. They are the building blocks of the more complex carbohydrates (polysaccharides) such as starch, cellulose, pectin, and glycogen (animal starch, stored in the liver). Sugars are also found in living things as parts of other organic molecules.

The complex carbohydrates don't taste sweet, but they are nutritious. The body must digest them or break them down into their building blocks before they can be absorbed as simple sugars. Carbohydrate digestion begins in the mouth, catalyzed by the enzyme amylase, secreted in saliva. Try this: Next time you eat a saltine cracker, hold it on your tongue until it begins to disintegrate. It will begin to taste sweet, because your salivary amylase has snipped some of the starch into its component sugars.

Today, sugar sweeteners are found in the diets of most people around the world, but it wasn't always so. The first real sweetener for humans was honey, a sticky syrup produced (and defended) by honeybees, so desirable that early humans risked the pain of stings to take it away from the bees, then began to figure out how to "domesticate" or manage the bees, which are extremely efficient in concentrating a widely scattered resource. To make a one-pound comb of honey, bees must collect nectar from about 2 million flowers.

The term *honey* has also been used to refer to the syrup of dates, sweet sap exuded by trees such as the European flowering ash (the manna ash), the boiled juice of grapes, or to the nectar stored by "honey" ants in the abdomens of individual ants that hang from the roof of the nest like clusters of grapes, living honeypots. For Australian Aborigines, the ants are a delicacy. But bee honey can be obtained in much greater volume.

As early as 2600 B.C., the Egyptians were keeping bees, as illustrated by tomb drawings. Domestic honey at that time may have been reserved for the rich and powerful, but beekeeping must have quickly spread to people of all stations, who valued it as a sweetener and for medicinal purposes or as a fermentation base for alcoholic beverages. Human languages are rich with references to honey as a metaphor for that which is pleasant or desirable. It's a widely used term of endearment. In Sanskrit, *madhu-kulya*, a "stream of honey," is used to express an overflowing abundance of good things. The Bible refers to Palestine as "flowing with milk and honey," a phrase with two sugar-containing substances, and the term has become a metaphor for an earthly paradise. The term *honeymoon* comes from the celebratory revelry catalyzed by drinking mead, a honey-based alcoholic beverage. (I have seen several explanations for the *moon* part, among them that the revelers drank for a month and that love wanes like the moon.)

Honey from honeybees contains primarily the monosaccharides fructose and glucose (table sugar contains these molecules, too, but in combined form, to make up the disaccharide sucrose). It also contains small amounts of proteins, vitamins, and minerals, plant perfumes, and perhaps plant toxins, depending on the nectar sources that the bees have used.

Humans are not the only honey lovers. Skunks, bears, and badgers are notorious for raiding beehives. Many insects, including ants, find honey a feeding paradise. A few years ago we kept a backyard hive, mostly to watch the bees come and go. We were

Honey is probably the oldest sweetener. Bees gather nectar from flowers, take it back to the hive, and concentrate it by fanning their wings over it. The honey, consisting of several sugars and a variety of other substances, is sealed in hexagonal combs constructed of wax flakes secreted by the bees.

gentlefolk beekeepers; after all, I could buy honey and sugar at the corner grocery store. I remember the first sweet sticky extraction day. I thought I had cleaned up, but I woke the next morning to find ants lined up at the kitchen countertop seams like cattle at feedyard trough, consuming the remains.

The love of sweetness may have contributed to the fall of the Roman empire. The Romans discovered that boiling grape juice in lead pans produced a sweet syrup, very sweet because it contained lead acetate, called "sugar of lead." They used the syrup as a flavoring. Taken together with the lead they consumed from drinking from vessels whose lead had been dissolved in wine, they poisoned themselves. Women became infertile, and over time the Roman upper classes weakened and died out.

Sugarcane (*Saccharum* spp.), a coarse sweet grass that contains sucrose, was known at least 2,000 years ago in the tropical islands of the South Pacific. People talk about new sources or kinds of sweetness (remember the flurry of discussion about rum-raisin ice cream?). Word got around. The Roman Pliny (A.D. 23–79) refers to a "kind of honey" extracted from bamboos. Sugar was probably introduced to Europe from India and brought by overland caravan route and by Crusaders returning from the East. Throughout the Middle Ages it was uncommon and costly, a luxury of rulers or a medicine sold in apothecary shops.

The grass is easily propagated in tropical climate, however, and in the fifteenth and sixteenth centuries the Portuguese and Spaniards spread it to the tropics of the world. Sugarcane then became one of the most demanded imports of Europe. The Arabs had developed a sugar-refining process in the fourteenth century. (In the New World, the North American Indians had long before devised a method of making sugar from the sap of the sugar maple tree.)

The beet, a temperate-climate plant, was known as early as 1590 to contain large quantities of sucrose, but the extraction process was not developed until 200 years later by a German chemist. Europe's sugarbeet industry got a boost when Napoleon cut off France's sugarcane import trade with England. The sugarbeet industry has also played an important role in the history of the western United States.

In the sucrose-refining process, raw sugar crystals, covered with an impurity-containing film, are an intermediate product. During

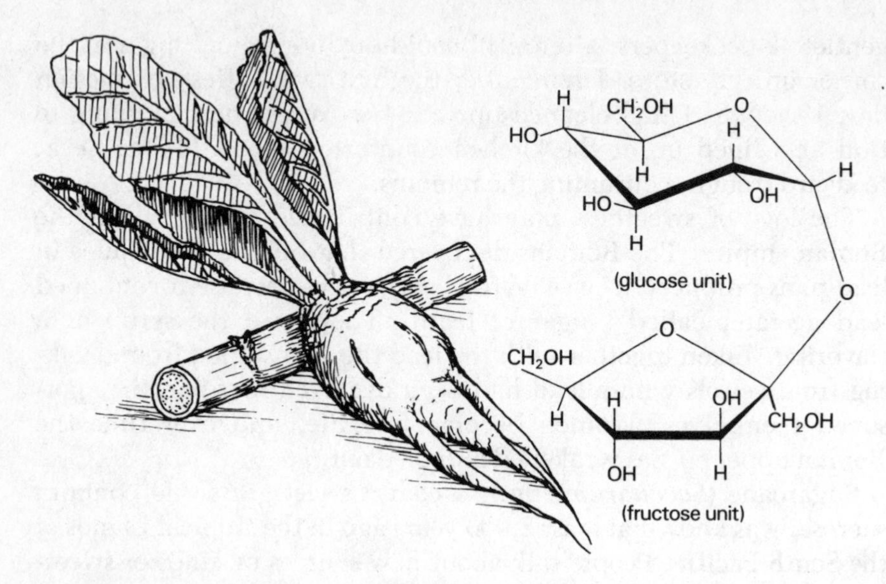

The tropical sugarcane and temperate sugarbeets are sources of sucrose, or table sugar. Sucrose consists of glucose and fructose sub-units.

further processing a dark-colored liquid is produced from chemical reactions that occur at high temperatures. This liquid, molasses, was commonly used in the 1700s and 1800s as a sweetener as well as a fermentation food for yeast in making rum. Molasses syrup contains 50 to 70 percent sucrose. The bitter by-product, blackstrap molasses, we feed to cattle.

Corn syrup was first developed in the 1920s, derived from corn starch by treatment of the starch slurry with acids, heat, and/or enzymes. The syrup, a mixture of molecules of different glucose chain lengths, is not as sweet as sucrose but is often used with sugar in candies, ice cream, and other food products because it adds texture and body. High-fructose corn syrup is a product of modern biotechnology. It was developed in 1970, made from an enzymatic process that converts the glucose derived from corn starch into fructose (fruit sugar), a simple sugar that has a higher perceived sweetness than sucrose. High-fructose corn syrup, a chemical concoction like high-octane gasoline, may contain 40 to 100 percent fructose. Because fructose tastes sweeter than sucrose, smaller amounts are needed for sweetening, resulting in lower-calorie foods.

High-fructose corn syrup is the primary nutritive sweetener in soft drinks. It is 30 to 40 percent cheaper than sucrose and is rapidly accounting for a large percentage of all caloric sweetener consumption in the United States.

Sugar as Mood Food

Consumption of carbohydrates stimulates the pancreas to produce insulin, which sets in motion a chemical chain reaction leading to the production of the brain neurotransmitter serotonin, which in turn produces a sense of calm and well-being.

Therein lies partially the appeal of chocolate sweets. Although a potato would do the same thing, chocolate tastes and smells better to most people, judging from consumption figures. In addition to the sugar added to the product, chocolate contains a perker-upper, theobromine, a relative of caffeine. And it contains a little phenylethylamine (PEA), an amphetamine-like brain chemical that we may produce when we fall in love. Scientists say that chocolate isn't addictive and that the PEA content isn't enough to make any difference, but there are those of us in love with chocolate. (Chocolate lovers are tipped to the female side. Apparently a male often prefers a beer to a candy bar.)

From 1964 to 1977 in the United States, smoking declined dramatically, paralleled by a rise in sugar consumption. While this can't be interpreted as a cause-and-effect relationship, recent studies indicate that carbohydrate consumption by abstinent smokers may be a kind of self-medication for tension, fatigue, and depression related to nicotine withdrawal. Others, with conditions unrelated to smoking, may use carbohdyrates in the same way, as a "mood food."

In one study, psychologists found that both cigarette-smoking humans and nicotine-injected rats chose fewer sweet foods. When nicotine was withdrawn, they both showed a preference for sweet carbohydrates. Also, humans who abstained from smoking were less anxious and hostile and twice as successful at staying off tobacco when they consumed a diet high in carbohydrates. Alcoholics are often helped through withdrawal with sweet desserts, and anecdotal evidence from methadone clinics suggests that former heroin addicts consume great quantities of sweet carbohydrates. The researchers suggest that the physiological mechanism underlying all sorts of drug dependence may be a neurochemical short-circuiting of the food pathways; the body may misinterpret certain foods—sugar in this case—as drugs. Other studies of obese patients who claim to have an irresistible craving for carbohydrates, as well as of patients with wintertime depression, suggest that these people load up on sugar and starch to alleviate fatigue and improve their moods.

Fooling the Tongue: Sugar Substitutes

Sweeteners make many foods more desirable, diluting the bitterness of chocolate, for instance, or the sourness of grapefruit. But sweeteners are blamed for contributing to obesity (a major correlate of adult-onset diabetes), dental caries, and hyperactivity of children, not to mention sugar highs and crashes. Sweeteners are all but impossible to give up, however. Our sweet tooth was adaptive in a primitive world. We're products of our evolutionary heritage in a world that now has an overabundance of sugar. What to do? Can we find a substitute that is sweet without the calories? Something that satisfies our tongues without making us overweight?

The sugar alcohols, such as xylitol, are useful in some ways as a sugar substitute. Most occur naturally in fruits and are chemically related to sugar. They add texture to some foods and are used as coatings on tablets and chewing gums and as sweeteners in gums and hard candies—items that are not consumed in large quantities by one person. The sugar alcohols are noncariogenic, that is, bacteria will not grow on them, so they do not promote tooth decay. Studies have shown a 30 percent reduction in dental caries in rats fed the sugar alcohols sorbitol and mannitol, in their diets and virtual elimination of caries in rats fed a xylitol-containing diet. Sorbitol, mannitol, and maltitol are considerably less sweet than sucrose, while xylitol has the same sweetness as sucrose. Sorbitol contains the same number of calories per gram as sugar sweeteners but is absorbed more slowly from the digestive tract than is sucrose, and so is useful in special diets. When consumed in large quantities, however, the sugar alcohols can have a laxative effect, as small children discover after they have eaten a whole pack of dietetic chewing gum. New sugar alcohols being developed in the laboratory may be less likely to produce diarrhea.

Saccharin, a derivative of toluene discovered in 1879 (quite by accident when a laboratory worker licked his fingers), is colorless, odorless, water-soluble, stable within a wide range of temperatures, and 200 to 400 times as sweet as sucrose. The body absorbs it slowly and excretes it unaltered. So it contributes taste but not calories. In the early 1900s saccharin was used in over-the-counter drugs and special foods for diabetics and was the only nonnutritive sweetener available in the United States for many years. It was useful in the two wars for folks at home, while soldiers were given real sugar. Slight alterations can change its taste from sweet to bitter.

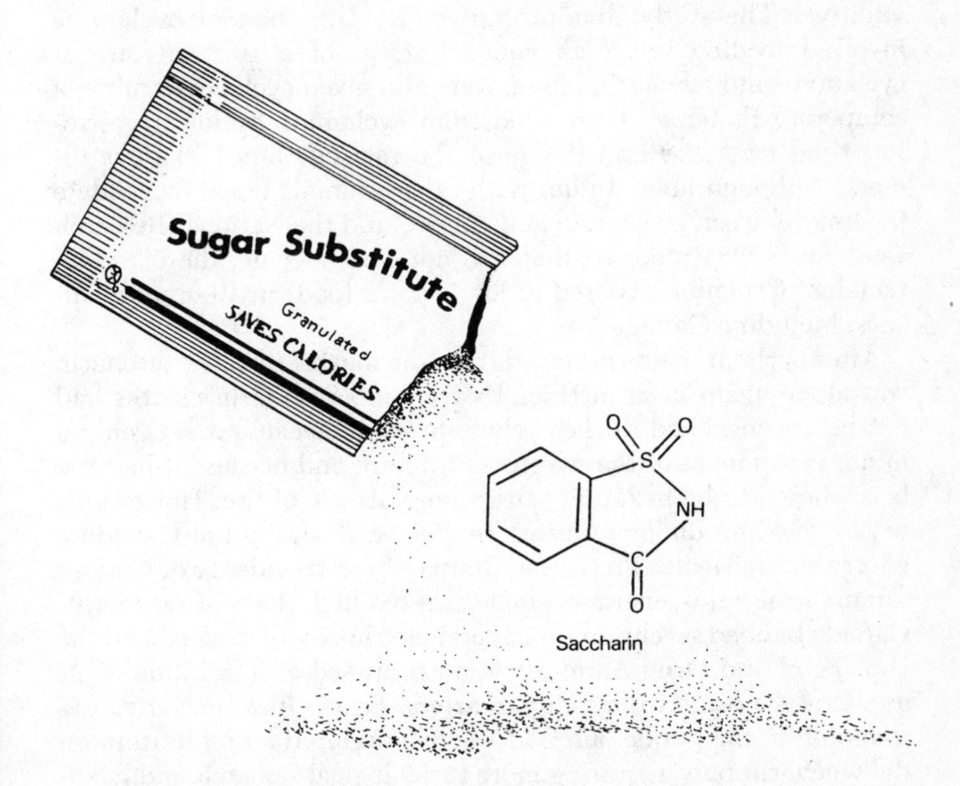

The nonnutritive sweetener saccharin was discovered more than a century ago. It was not used widely in the United States until the 1950s, when thinness came into vogue. Slight alterations can change its taste from sweet to bitter to tasteless.

Cyclamate, 30 times sweeter than sucrose and also heat-stable, was discovered in 1937, also accidentally—a laboratory worker had momentarily rested his cigarette on a laboratory bench contaminated with cyclamate. It was first marketed as a dietetic aid in the early 1950s and became a popular sweetener in diet beverages in the 1960s. I remember drinking cyclamate-sweetened Kool Aid in front of the television set on July 20, 1969, watching the astronauts land on the moon. Cyclamate doesn't produce an aftertaste, and it overcomes the bitterness of saccharin, and so was commonly used with saccharin in a 10:1 combination, which resulted in greater sweetness than either substance alone, a synergistic effect seen in combinations of other sweeteners.

Cyclamate is now banned from use in the United States as a food additive. The study that prompted the U.S. ban of cyclamate involved feeding rats high concentrations of a 10:1 mixture of cyclamate and saccharin. Some were also given cyclohexylamine, a compound that may be formed from cyclamate by lower gastrointestinal tract bacteria. Some of the rats developed bladder tumors. Although later studies with other animals failed to validate the link between cyclamate and cancer, and the National Research Council in 1984 reported that it is not carcinogenic, the U.S. ban remains. Cyclamate is used in low-calorie foods in 40 other countries, including Canada.

After cyclamate was removed from the market in 1969, saccharin was alone again as an artificial sweetener. Saccharin's status had not been considered in the cyclamate study because it was only a minor component of the sweetener mixture and because it had not been shown to be hazardous in its long history of use. However, it began to come under scrutiny. In 1977 a study by the Canadian government's Health Protection Branch showed evidence of bladder tumors in second-generation male rats fed high doses of saccharin. Canada banned saccharin for general use shortly thereafter, and the U.S. Food and Drug Administration proposed a prohibition of its use. Protests came from many quarters. People like saccharin. Responding to the public outcry, Congress declared a moratorium on the saccharin ban, requiring more toxicological research and labeling of saccharin-containing food: "Use of this product may be hazardous to your health. This product contains saccharin which has been determined to cause cancer in laboratory animals." Studies so far have not established a health risk to humans from the use of saccharin in normal dietary amounts; however, saccharin's potential effects on children and pregnant women have not been well studied. The American Medical Association supports the continued availability of saccharin, suggesting that consumption patterns and reports of adverse health effects should continue to be monitored carefully. Saccharin is approved for use in more than 80 countries.

In 1965 aspartame was discovered by researchers looking for an inhibitor of the hormone gastrin as a possible treatment for ulcers. Aspartame was an intermediate in the process. It is a methyl ester of two amino acids, phenylalanine and aspartic acid, both found naturally in food. To purify it, it was heated in a flask. The mixture bumped to the outside of the flask. Later, the laboratory worker who had handled the flask licked his fingers to pick up a piece of paper and noticed a strong sweet taste. Another accidental discovery.

The aspartame sweetness could not have been predicted. Aspartic acid is tasteless to sour, and phenylalanine is bitter. It enhances some flavors and in combination with saccharin again results in a sweeter taste than either one has alone. Its best use seems to be in acid-containing beverages. It loses flavor in dairy drinks, and it degrades at high temperatures so is not used in canned or cooked foods. Proteins are not durable chemicals when heated—note how an egg changes when it is cooked. People who have the genetic disorder phenylketonuria and cannot metabolize phenylalanine should not consume aspartame.

Aspartame, which is sold under the brand name NutraSweet, has also been embroiled in controversy. The Food and Drug Administration has received several thousand complaints from consumers, who claim that aspartame-containing foods have caused illnesses ranging from diarrhea and nausea to dizziness, mood changes, headaches, epileptic-like seizures, and menstrual irregularities. Research has not clearly linked these symptoms to human aspartame use, and several new studies seem to have reaffirmed its safety.

In the United States in 1985, the estimated total per capita consumption of all nutritive sweeteners was about 127.4 pounds. Consumption of aspartame and saccharin has more than doubled

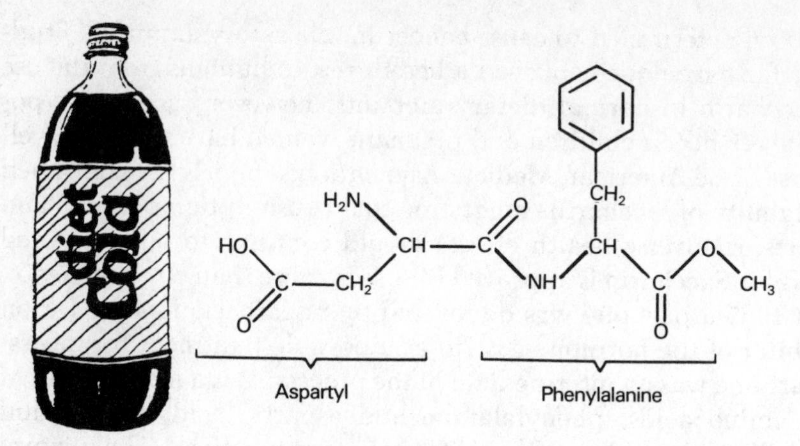

Aspartame, a protein sweetener, was discovered in 1965. It is digested in the same way as other amino acids and is calorically equivalent to sucrose but is about 200 times sweeter, so only a very small amount is needed to sweeten products.

since 1978, but in 1984 it was only about one-half ounce per person. The sweetness equivalency of the low-calorie sweeteners related to sugars (1/200) allows us to consume much smaller amounts.

Several other sweeteners of natural origin are used to sweeten foods in other countries. Most come from plants that do not grow in the United States: glycyrrhizin from licorice roots from Europe and central Asia; stevioside from the leaves of *Stevia rebaudiana*, Paraguay; and thaumatin I and II from the fruit of katemfe, *Thaumatoccus danielli*, monellin from the serendipity berry, *Discoreophyllum cumminsii*, miraculin from the miracle fruit *Richardella dulcifiea*, all West Africa. Glycyrrhizin and thaumatin have been approved for limited use in the United States.

Most sweet-tasting molecules are fairly small, but thaumatin and monellin are large protein molecules, and they are exceptionally sweet: 100,000 times sweeter than sugar. Thaumatin is a single-chain protein made up of 207 amino acids, and monellin is made of two chains, one with 44 amino acids, the other with 50 amino acids. Chemist Sung-Hou Kim and his colleagues at the University of California at Berkeley and at Columbia University have figured out the complicated three-dimensional structure of thaumatin and monellin and have tried to determine what makes them so intensely sweet. Despite the two proteins' resemblance in taste, the researchers reported, they have no significant similarities in their crystal structures or amino acid sequences. The researchers are trying to determine which structures are responsible for the sweet taste receptor binding. These proteins, they say, can be a starting point for engineering of protein sweeteners with "better physical and chemical properties than their natural counterparts." The search goes on.

Among the vertebrate animals, scientists have discovered a variety of intriguing taste responses, indicating different taste receptors or different processing. Some are clearly related to what is important in the animal's diet; others are puzzling. Pigeons don't taste sugar. Chickens and cats don't taste either sugar or saccharin. Dogs reject saccharin. Rodents, which are closer to us in their eclectic tastes, seem to like saccharin but don't like cyclamate. Squirrel monkeys reject saccharin and cyclamate but accept dulcin, another substance sweet to humans. Gymnemic acid, which blocks the sweet response in humans, has the same effect in dogs, hamsters, and macaques but not in rats or squirrel monkeys.

Sour, Salty

Acids produce a sour taste. They also produce a hydrogen ion when they are dissolved in water, which is thought to be responsible for producing sourness. However, different acids produce different tastes, and they are affected by other chemicals, including salts in the saliva, which buffer them. Acids and sugars mixed together seem to interact in a complex way, so that sweetness is related to acidity.

Citric acid combined with sucrose in the right proportions, plus a chemical for odor—such as (the Eskimo-sounding) nootkatone, found in grapefruit oil—will produce a basic fruit flavor. Fruits contain various acids in various proportions. Citric acid is an important constituent of berries, citrus fruits, and pineapple. Tartaric acid is a major acid in grapes. Malic acid contributes to the taste of apples, pears, cherries, and apricots. In addition, the fruits contain different sugars in different proportions. The sugar and acid makeup of fruits change as fruits progress from unripe to ripe to overripe, and our tongues tell us about it.

Ascorbic acid, or vitamin C, is found in citrus fruits, tomatoes, and green vegetables. Its exact role is not understood, but it promotes many metabolic functions, particularly protein metabolism. It works with antibodies to promote wound healing, and as a coenzyme it may combine with poisons, rendering them harmless, until they are excreted. Some animals can synthesize vitamin C in their bodies, but humans must obtain it from dietary sources. Vitamin C is water-soluble; excesses are not stored by the body, so a continuous supply is necessary. Scurvy, identified as a vitamin C deficiency disease only in this century, was an expected, if not understood, risk of every long sea voyage, where sailors' rations represented anything but a balanced diet. (Magellan, on his voyage in the 1520s, escaped scurvy as a result of the quince jelly he kept in his captain's pantry to relieve the monotony of the diet.) In the United States, in winterbound frontier communities, spruce beer was sometimes recommended as a treatment (the inner bark of spruce trees contains ascorbic acid). Fresh fruits and vegetables must have looked and tasted awfully good to the sailors in port or to the frontier people in the spring.

Appetite and the chemical senses have an interactive relationship. The flavor of foods affects the desire to eat and guides food selection—the *hedonic response*—while changes in hunger and satiety affect reactions to the taste and smell of food. That is, when you're full, food isn't as tempting as when you're hungry. It's the grocery-shopping maxim: I always buy more at the grocery store when I shop hungry.

Changes in metabolism may alter food intake by affecting the hedonic response to food. The pleasantness of sour and salt are often determined by the momentary state of nutrition of the body. This seems to be true to a lesser extent for the sweet taste. The lack of a nutrient seems to intensify taste sensations, resulting in people choosing foods with the taste characteristics of the deficient nutrient.

The craving for and compulsion to eat nonnutritious or unnatural foods, such as dirt (especially clay) or laundry starch, is called *pica*. The disorder of pica is documented all over the world and was known at least as far back as 40 B.C. Several explanations have been suggested, including a kaopectate effect or a need for mineral micronutrients. It has been related in some cases to iron or zinc deficiency, but victims of pica don't seem to eat substances that are high in iron.

Substances that taste salty are those that dissociate in water into positively and negatively charged ions, cations and anions, respectively. Sodium is a cation, for instance; chloride an anion. Cations may excite taste receptors, and anions generally inhibit them. Sodium chloride (table salt), sodium bromide, and sodium nitrate have the sodium cation in common and produce sensations of the same magnitude. As the size of the anion increases, however, salts with a common cation produce responses of a decreasing magnitude. Salts with the anion in common—sodium chloride, magnesium chloride, ammonium chloride, potassium chloride—produce neural responses of different magnitudes. None except sodium chloride tastes very salty, and ammonium chloride and potassium chloride are not efficient in stimulating the fiber that responds to sodium chloride.

Salts in the bodies of living things play crucial roles in the passive control of osmosis, the movement of water into and out of cells. An isotonic solution, such as my contact-lens saline solution, pulls evenly. It contains the same salt concentration as body cells and so is comfortable to put in my eye. Intravenous fluids administered at the hospital to dehydrated people are also isotonic solutions. Drinking a hypertonic salt solution, such as seawater, is counterproductive, even deadly, for humans, because it pulls water out of cells, dehydrating them, as the water passively equalizes its concentration inside and outside cells. (The reason that gargling with salt water will shrink a sore and swollen throat.) Water follows salts, and the body's cells work actively, use energy, to keep the correct salt balance for life. (Some animals, such as seabirds, can drink seawater, but they have special nasal glands to excrete the excess salt.)

Animals will seek out and preferentially take in food and water with a high salt content when their diet has been low in sodium. I have a photograph of my bare feet on the red sodium-poor soils of Misiones Province, Argentina. On my toes are butterflies, proboscises extended, sucking up the salt I had lost through perspiration. Animal studies indicate that salt intake seems to be an innate response regulated by the size of the deficit. Animals will lose interest in a sodium solution when they have consumed enough to meet the deficit. Studies of sheep and goats have shown some kind

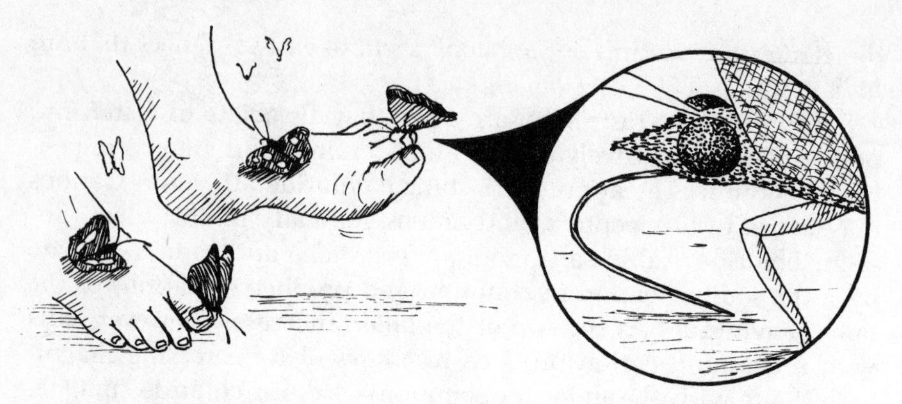

A butterfly laps up human perspiration, a source of salt. Salts are essential in producing the electrolytic properties of animal cells, and a proper balance is necessary for nervous transmission and muscle action.

of intermediary regulation must take place: They will stop consuming sodium 15 to 30 minutes before the sodium is absorbed and reaches optimum levels in their blood.

Laborers from swelteringly hot steel mills and coal mines have been known to flavor their beer and food liberally with salt after their work shifts. This made their food unpalatable to everyone else, even to them if it was done at any time other than immediately after work.

Hunger for salt has also been reported clinically: One young boy with adrenal gland deficiency, which can result in rapid salt depletion, survived by consuming table salt by the handful but died when his salt intake was restricted.

A strong dose of salt water is the oldest way known to induce vomiting. In normal people, unpalatable concentrations of salt are often metabolically toxic, but salty tastes are usually beneficial if they taste palatable. Newborns do not prefer salt water. The ability to perceive sodium relates to the development of the taste receptor for it; it develops in infants at about four months, at a time when they are unlikely to have been exposed to salty-tasting foods or beverages (except perhaps a sweaty nipple?), ruling out a learned response.

I happen to think we like salt because it tastes like ourselves, which could be expanded either to sex or hunger contexts (the taste of a sweaty man—or woman, I suppose—or the taste of fresh animal food, because other animals are salty too). I wouldn't drink my contact-lens solution, but when it drips down my cheek and I lick it with my tongue, it is not unpleasant.

Humans (and some other animals) seem also to have a need-free appetite for salt; hypertension, or high blood pressure, may be caused or exacerbated by excess sodium intake. Where did the appetite come from? It's the subject of some disagreement and seems to be complicated by a variety of factors. A specific physiological appetite may tell us we want salt, the same as it tells us we want water or calories. But a range of salt consumption can be tolerated, and healthy people can excrete excess salt. Human salt taste perceptions and preferences have been found to change with the amount of salt in the diet. Decreasing dietary salt is followed, within a month or two, by a shift downward in the level of dietary salt found optimal; a parallel shift in the opposite direction occurs following increased salt intake. This is good news for those who need to reduce salt intake. It may be explained by something as simple as changing the body's set point for salt and a change in metabolic handling of it.

Some think that humans evolved on a diet that contained adequate amounts of sodium. Because the kidneys and gut are fast and efficient in regulating sodium, and sodium is also stored in bone, they argue, perhaps the need-free appetite for salt is an appetite that evolved when evironmental sodium was scarce or to protect us from sodium depletion. Others suggest that the evolutionary diet may have been high in potassium and that extra sodium was needed for potassium excretion; however, animal experiments in which dietary potassium was increased did not result in an increased salt appetite. Still others suggest that salt hunger may be part of the adaptation of tropical humans to a temperate environment, that humans added salt to the "unpalatable pap" they began eating with agriculture, even that it is one of humankind's oldest addictions.

Like sugar, salt has not always been eaten in such abundance. It must have been unattainable by primitive humans in many parts of the world. Homer's *Odyssey* speaks of inlanders who do not know the sea and use no salt with their food. Homer calls salt "divine" and Plato calls it "a substance dear to the gods." The ancient Germans waged war for saline streams and believed that the presence of salt in the soil made a district a place where prayers were more readily heard. In the early history of the Roman army and in later times an allowance of salt was made for officers and men. In imperial times, this *salarium* (from which "salary" was derived) was an allowance of money for salt. Salt was also important as money in Africa and Asia. Some of the earliest trade routes were created for traffic in salt. In some parts of America, salt was first introduced by Europeans.

Salt's value does not just lie in adding flavor to bland foods. It has been important for centuries as a preservative. A lover of French cooking remarked to me that Americans have deadened their tastes for the delicate flavors of fine foods by their practice of oversalting and by their practice of preserving foods in salt and in sugar.

Bitter, Hot, Cold, and New

Humans have a natural aversion to bitter, just as they have a natural affinity for sweet. The bitter aversion is found throughout the animal kingdom, from protozoa and jellyfish to birds and mammals. The bitter substances are generally secondary compounds of plants, compounds that take more energy to make and store than do the primary compounds of carbohydrates, proteins, and lipids. The secondary compounds have evolved as a defensive mechanism to protect plants from being eaten or from invasion by organisms that cause disease.

Many animals, especially amphibians and certain insects, also produce bitter toxins. One I'm familiar with is the giant water bug that I spent a lot of time collecting in Arizona streams. Water bugs are also called toe biters, so they have another line of defense, but they are big bugs, and one represents a nice morsel for a fish or a bird. When I picked them up (carefully, grasping them from the sides), they sometimes exuded a milky-blue bitter substance from glands between the thorax and abdomen. (I licked it once, like the careless laboratory workers.) The warty glands of toads produce poisonous bitter alkaloids. A dog that mouths a toad won't do it again. The bitter poisons work their strategy on animals that learn.

The bitter compounds are poisonous alkaloids, terpenes, phenolics, or glycosides that are stored in plant tissues or exuded to deter the plant's enemies. They are especially effective against bacteria, fungi, and insects. More than 4,000 alkaloids have been identified. Among the best known to us are caffeine, quinine, codeine, nicotine, morphine, and strychnine. Digitalis, produced by the foxglove, is a glycoside that affects the heart muscle of vertebrates. But aren't these chemicals drugs? And don't many of them become addictive? Do some humans overdo them, just as they eat too much sugar and too much salt?

As is the case with many poisons, they may be therapeutic in small amounts, but they can also disrupt normal biological functions to the point of death. We may consume a small amount, a sublethal dose (even a sublethal dose of several compounds) that nevertheless discourages microbial pathogens and parasites. But too much is not good for us. Very bitter substances detected by the taste buds on the back of the tongue will evoke a gagging reflex, with the adaptive significance of getting rid of a harmful substance. Other mammals also differentiate between high and low alkaloid variants. Humans have another way of getting around the poisons: cooking, which destroys them.

For humans, a spoonful of sugar helps the medicine go down. My mother sweetened the paregoric that she gave me for stomachaches, and Michael the pharmacist tells me that drugs, often alkaloids in the water-soluble "salt" form, commonly are mixed with sugar binders or coated with sugar or placed within a gelatin capsule. Tranquilizers and steroid drugs generally taste sweet; no problem there. Pharmacists, Michael says, use the term *bitters* for quinine drugs. Quinine, from the bark of the *Cinchona* plant of the northern Andes, means death to the active phase of the malarial blood parasite. Actually, four main alkaloids are found in *Cinchona* bark, making up from 7 to 17 percent (in selected plants) crystallizable alkaloids. Quinine-containing drinks, such as tonic water or bitters, seem to be an adult, acquired taste. But then, the drinks have a mixed profile. A gin and tonic, for instance, is also sweet and sour.

Humans like to live a little on the edge with their senses. According to a scientist who has studied how people gradually learn to eat chili peppers, humans are the only animals that are thrill seekers and knowingly incur limited risks, among them riding roller coasters, taking very hot baths, and eating horseradish, ginger, and hot chili peppers, which are actually oral irritants, rather than substances that can be classified in the four-quality scheme.

Mexican children aren't born liking chili peppers. They acquire the taste. The capsaicinoids that impart the flaming bite to chilies have almost no odor or flavor. They act directly on the pain receptors in the mouth and throat and will irritate other tissues as well. One drop of capsaicin in 100,000 drops of water will produce a persistent burning on the tongue. Capsaicin makes warm solutions appear hotter and cool solutions less cool. It also lessens perception of sour and bitter tastes, while piperine, in black and white pepper, affects the sweet and salty tastes as well. Curry will also anesthetize the tongue. Menthol, which occurs naturally in peppermint, influences the thermal senses, producing long-term enhancement of perceptions of cold. (A little mentholatum in a stuffy nose doesn't unstuff it. The cold sensation tricks the brain into thinking it does.)

Seasoned adults also like new taste sensations. I love going to dinner at someone else's house and eating something different. I live in a small city. We have good Italian, Chinese, and American eateries. A restaurant featuring Cajun cooking opened recently. It was swamped with customers. The omnivore rats, again, are not so different from humans. Studies have shown that when rats are presented with a palatable, novel diet they increase the size of their first meal on this diet.

With the exception of the tip of the tongue, the mouth is unusually insensitive to warming, including warming in the range of temperatures that eventually burn the tissue. The mouth's thermal sensitivity is not the same as that of the external skin, but our tongues have come to us from ancestors who only comparatively recently discovered fire and began to cook their food. Liquids as much as 4 degrees Centigrade below body temperature are perceived as thermally neutral in the mouth. A liquid painfully hot to the fingertip (45 degrees Centigrade) is felt as only mildly hot in the mouth.

Fats are important in providing texture and "mouth-feel" to foods. They are also high in caloric content, with about twice as many calories as the same amount of proteins or sugars. (This is why fat is such a good storage chemical for animals. They can convert extra calories, regardless of the form in which they were taken in, into fat and not have as much weight to carry around.) But excessive consumption of fats has been implicated in coronary heart disease, cancer, diabetes, obesity, and hypertension. The tendency toward fat consumption is probably a result of our evolutionary history when availability of food or the efficient use of it determined survival. People with the thrifty gene were more likely to survive. High-fat foods were generally available only at times of a feast, but we can now feast three times a day, so food scientists are looking for artificial fats. One, introduced recently, consists of proteins that have been processed through heating and microparticulated into beads one one-thousandth of a millimeter across. The protein beads give the tongue the same sensory stimulation as fats. The artificial fat has 1.3 calories per gram, compared with regular fat's 9. It cannot be used in cooking because it is not heat-stable, but it is likely to be used in diet ice cream and salad dressings.

Differences of Tongues: Season to Taste

My friend Jeanne tells me that artichokes have an extremely pleasant aftertaste. "It's clean," she says, when asked to describe it. "It's not sweet or salty. It's refreshing. It lasts 30 to 45 minutes. The artichokes have to be freshly cooked. Marinated or highly seasoned artichokes don't do it." Chris says the artichoke aftertaste is "like the smell of rain in the desert."

I'm mystified. I like artichokes, but I've never experienced the aftertaste. I feel a little cheated, a little disappointed in my artichoke-sensing abilities and in our differences in tongues. But tongues, and tastes, are different.

A chemical known as n-phenylthiocarbamide (PTC) is one of a group of about 40 related compounds used as antithyroid drugs. PTC tastes extremely bitter to most people. In 1931, however, Arthur L. Fox reported that many other people find it tasteless. This bimodal (either/or) picture of taste piqued the interest of human population geneticists. The inheritance of ability to taste PTC and the PTC-like chemicals depends on a single pair of genes, they found. Tasters have the dominant gene, which probably codes for a receptor protein, in a single dose (heterozygous) or a double dose (homozygous). Nontasters have a pair of recessive genes. Mice also show some bimodal distribution of taste sensitivity.

Barium sulfate emulsions, drunk by people before their gastrointestinal tracts are X-rayed, and creatine, found in lean meat, are also tasteless to some people but bitter to others.

In genetics laboratory exercises, taste-testing of PTC and its relative thiocarbamide is done with little pieces of filter paper impregnated with chemicals and a nontreated piece of filter paper as a control. Students who don't taste the substances roll the filter paper around in their mouths, looking with puzzled amusement at their grimacing classmates who have spit out the paper.

Early studies showed different populations have differing frequencies for the PTC-tasting genes. In the North American white population, for example, 20 percent were homozygous PTC-tasters, 50 percent heterozygous tasters, and 30 percent nontasters. Hindus had a higher frequency of tasters, and the Japanese and Brazilians had a lower frequency of tasters. Further studies of threshold levels showed that expression of the trait is more complicated than just tasting or not tasting. Some people are sensitive to very small amounts, and a few others may taste PTC as sweet, sour, or salty. Researchers also looked at taste reactions to thiocarbamide (thiouracil) and found that both PTC tasters and PTC nontasters detected it; many reported it to taste "nauseating."

Sodium benzoate, sometimes used as a food preservative, is another substance tasted by some people and tasteless to others. In one study that involved testing people for sensitivity to both PTC and sodium benzoate, every combination was found except tasteless for PTC and bitter for sodium benzoate. The most frequent combinations in descending order for PTC/sodium benzoate were: bitter/salty, bitter/sweet, bitter/bitter, and tasteless/salty. These taste abilities may also be correlated with food likes and dislikes. The bitter/salty people, those who found PTC bitter and sodium benzoate salty, were the ones who liked "controversial" foods such as sauerkraut, buttermilk, spinach, and turnips, whereas the bitter/bitter tasters were averse to them.

In another study, 175 college students were given a list of 120 common foods and checked off their food dislikes. Some students liked all the foods, others disliked more than half of them. They were tested in other ways, including for their quinine taste thresholds. The study found a correlation between a student's percentage of food dislikes and the threshold for detecting quinine. As the threshold decreases—as they can taste it in smaller and smaller dilutions—the percentage of food dislikes increases. (Women have also been found to have more food taste dislikes than men.)

For the 25 percent of students who had high taste thresholds for quinine, food dislikes could not be predicted reliably; the 25 percent with low thresholds disliked many foods. For the 50 percent in the middle, cultural, social, and idiosyncratic variables seem to decide food likes and dislikes. The researchers found that sensitive quinine responders, who were more particular in their choices of food, were also more likely to be nonsmokers. Heavy cigarette smokers were more likely to be found among the insensitive quinine taste responders.

Difference in sensitivity to substances such as quinine is a reflection of a more general systemic sensitivity. For example, a sensitive taster, one who needs fewer quinine molecules in water to differentiate the solution from water alone, will also need fewer molecules of a drug to obtain its pharmacological effect. Researcher Roland Fischer examined the taste thresholds in a group of acutely anxious mental patients who developed Parkinsonian tremor as a side effect of certain tranquilizers. He found that the taste-sensitive paients were also drug-sensitive and always reacted to the drug more quickly. They required two-and-a-half times smaller cumulative doses of the tranquilizer trifluoperazine than taste-insensitive patients. Sensitivities to other water-soluble drugs have been studied in the same way with similar results. Fischer concluded that this observed relationship between taste threshold and drug reactivity should enable doctors to prescribe individualized therapeutic doses of a drug instead of mechanically vending drugs by weight. This is probably happening in some quarters, although not correlated with taste: I know a psychiatrist who prescribes psychoactive drugs such as tranquilizers and antidepressants, then does periodic "med-checks" to talk with patients about the effects and to fine-tune dosages.

Several studies have indicated that people seek out mates who resemble them in appearance. (Do your own study. Look at the photographs of couples on the wedding page of your newspaper, if your newspaper still runs them: face shapes, eye shapes, smiles. Couples look even more alike 50 years later. Check out the photographs of those who are having their golden anniversaries.) Fischer found that people seem to seek out partners and companions with the same tastes: tongue tastes, not tastes in china patterns. Seventeen out of 20 married couples Fischer tested in the 1960s had similar quinine taste thresholds, and nearly all of 40 pairs of good friends had the same range of taste sensitivity. He didn't know quite what to make of it then—was his sample skewed? But since that time, the further collection of information about relationships between taste sensitivity and drug sensitivity, personality traits, cortical arousal, reaction times, food dislikes, and smoking habits suggest that sensitive people prefer other sensitive people, and insensitive people prefer insensitives. Fischer suggests that some partners may come closer together in this respect over time too; that is, sharing of a sensitive taster's living habits by a less sensitive taster might result in a lowering of taste thresholds. Social factors may also have to be considered. If your mate doesn't like Mexican food, it's no fun. But shouldn't this be sorted out beforehand? If a guy doesn't like Mexican food (or green peas), I lose interest fast.

The perception of PTC has been reported to decrease with age. However, when only nonsmokers were tested, no significant age- or sex-related differences in taste sensitivity for quinine or a PTC-related compound was found. Among heavy smokers, a significant deterioration in taste responsiveness was apparent with increasing age. Aging and smoking together seem to exert a progressive reduction in taste responsiveness.

The number of taste buds decreases with age. L. L. Cudmore writes that at age 75 we have 36 percent of the taste buds we had at age 30. Dentures also cover up extralingual taste buds and help explain my father's penchant for putting pepper and hot sauce on everything.

The deterioration of taste and olfactory acuity among the elderly may present one of the major hurdles to the new food product developer. Because older people seem to have decreased sweet and salt sensitivity, they perceive as bland food that younger people think is just right. Bitter and sour notes in foods may be perceived as enhanced by older consumers.

Foods designed to be very sweet, such as desserts and some breakfast cereals, may find their appeal strongest at both ends of the continuum: children and the elderly. However, because the elderly have a high incidence of diabetes, they should probably avoid foods with excessive sugar concentrations. Enter the artificial sweetener.

The Fine Art of Tasting

Tasters can sharpen their senses, just as piano tuners do. My friend Larry is a meteorologist, a dedicated runner, and an oenophile—a wine lover. Larry is a disciplined, sensitive, temperate man, and his appreciation of the drink made from fermented grapes is scientific and artistic. He appreciates the art of the labels, getting a bargain price on a good vintage, cellaring wine to its optimum age, and—most of all—he takes sensory delight in consumption of the ancient beverage that Socrates spoke of: "If we drink temperately, and small draughts at a time, the wine distills upon our lungs like sweetest morning dew."

More than 50 substances have been identified as contributors to the flavor of wine, and there are undoubtedly a great many more.

A certain ritual is necessary to perceive all the characteristics—appearance, bouquet, taste, and aftertaste—of a wine, and the ritual begins by preparation for tasting, which includes avoidance of cocktails and such foods as raw onion, garlic, mustard, peppers, curry, and vinegar. The taste buds should be fresh, not fatigued.

The wine is also prepared for performing: Red wines should be permitted to breathe with the corks removed from the bottles for a period of time before tasting. If the red wine is very old, however, one does not want to miss the ephemeral bouquet that disappears quickly, like a pent-up genie escaping from the bottle. White wine must be chilled. Clear glass and a white tablecloth will aid the visual inspection. Look at the wine. It should be clear, although reds may have sediment collected in the bottom of the bottle; if so, decant carefully. Reds are ruby, although old reds may be tinged with brown. A wine that is overwhelmingly brown is not good. Whites may have golden highlights. Swish the wine around in the glass. The legs, or rivulets that run down the sides of the glass, are an indication of the glycerin content, a natural component of a balanced wine.

Then, the bouquet, or aroma, should be pleasant. It may be subtle or it may be strong and flowery, in which case it is called a nose. In great wines the bouquet is complex and challenging, evoking images of such things as herbs, spices, violets, lilacs.

To assure that the taste buds will be fully exposed to the wine, hold the wine in your mouth for a few seconds, sloshing it around, drawing in air, almost gargling. Expert tasters tilt their heads back slightly and make gurgling sounds as they aerate the wine. The air seems to accentuate the taste, they say, and aids identification of its various characteristics. The characteristics of smoothness and body are related to the wine's density.

The basic qualities of sugar and acid enter into the description. A dry wine, for instance, is lacking in sweetness. A fruity wine provides a pronounced taste of the grape. *Tannic* refers to the presence of astringent tannic acids. The vocabulary is also olfactory: A red Burgundy may suggest violets or lilacs or mint, while the taste of a rich Pomerol produced mostly from the Merlot grape may remind the taster of the smell of fresh road tar. Wines are described as smooth, delicate, or full-bodied, structured, closed, opulent, lush. *Dusty* refers to the impression that dustlike particles of flavor can be detected on the tongue; *earthy* to a presumed detection of the soil on which the grapes were grown; *nutty* to a spicy taste; *woody* to a woody taste; *foxy* to the taste of wild grapes; *flinty* to a slight metallic taste.

Swallow. The tasting experience is not complete without the aftertaste. The impression that remains may differ substantially from the impression given by the bouquet and taste of the wine in the mouth. Acidity may become more evident in the aftertaste.

Wine tasters clear their palates by rinsing the mouth with water or by eating a little bread. Cheeses go nicely with wine, but if you're just after the wine-tasting experience, stay away from strong-flavored cheeses because their flavors compete and affect the tasting experience. The wine merchants say, "Sell on cheese, buy on bread."

Novice tasters may at first detect only two dimensions: sweet-dry, heavy-light. But with practice they can improve. To develop a finesse in winetasting, many wines must be tasted and compared. Balanced wines are the ideal. But balanced for what? Balanced, of course, for an individual's particular palate. My wine book says this: "Tasting is a completely subjective experience, and each person will react differently to a particular wine. . . . Nobody has the right to dictate how you should react to a taste sensation."

Chemistry and even techniques such as laser fingerprinting have been applied to characterize wine, and they may improve vinting, but human tastes will be the ultimate judge.

Professional tasting in all areas takes dedication and concentration, but most people can learn. At Utah State University's food and nutrition laboratories, where many new dairy products are tested and tasted, taste panels are trained by first tasting foods with exaggerated defects. In milk, for instance, the tasters may detect traces of the forage the cows ate, "cooked" tastes imparted by the heating process, and oxidized chemicals. They rate the tastes of dairy foods on several scales but also have a strictly hedonic scale, a measure of how much they like or dislike them, which takes into consideration differences in such untastables as body and texture.

Taste Disorders: Dysgeusia

Dysgeusia is a general name for a taste dysfunction. *Ageusia*, or *hypogeusia*, refers to loss or diminishment of taste; *pantogeusia* to a constant bitter taste, unrelated to eating; and *aglycogeusia* to absence of sensitivity to sugar. Dysgeusia may be caused by such things as poor oral hygiene, gum disease, hepatitis, or pregnancy. Medicines, especially those that are secreted into the saliva, can also affect taste: Sodium pentathol and antidepressant drugs will often cause a metallic taste that lingers in the mouth, for instance; chemotherapeutic drugs taken by cancer patients may cause a taste dysfunction and contribute to a patient's loss of appetite. Dysgeusia may also be neurological. If a bad taste is localized on a specific area of the tongue, the problem may indicate damage to one of the nerves leading from the tongue or brain damage from epilepsy, a stroke, or a brain tumor.

Historically, loss of taste is probably just as common an affliction as loss of sight or hearing, but not as much is known about it because its recognition is recent.

Many people with taste dysfunctions have been misdiagnosed as having mental problems and have been sent to psychiatrists or psychologists. Robert I. Henkin, a physician at the Center for Molecular Nutrition and Sensory Disorders at Georgetown University Medical Center, says that unlike vision or hearing, taste has not been of great interest to medicine, probably because its pathology never seemed to have much diagnostic value. That is changing.

One of Henkin's patients was Adolph (Rudy) Coniglio, perhaps the first victim of idiopathic dysgeusia—taste disorder of unknown origin—whose experience has been recorded in detail. Rudy was 53 and the proprietor of a pizza restaurant in New Jersey in 1969 when he had an attack of flu and his chemosenses went on the fritz. Food smelled and tasted like garbage, like burned plastic, he said. The smell of grass, of his pillow, made him sick. Cigarettes, coffee, and bananas were repulsive. The only environment he found tolerable was the woods behind his house. He survived on the blandest food: a little cold milk, a little cold boiled potato, a little vanilla ice cream, a white grape. They didn't taste good, but they didn't taste as bad as other things. Doctors gave him antibiotics, tranquilizers, referred him to a psychiatrist, thought he had a tumor, ruled it out. Finally Rudy met up with Henkin.

At that time, Henkin had seen two cases of idiopathic dysgeusia: in an Air Force colonel who had been living on lettuce, cottage cheese, milk, and ice cream; and in an English colleague whose eating patterns were much the same. On a hunch, he had treated them with zinc sulphate tablets. He knew that D-penicillamine, used to treat various diseases, can diminish or distort taste and that administration of the drug is followed by a drop in the concentration of copper in the blood. When the patients whose taste distortions were associated with penicillamine were treated with copper sulphate, normal taste returned. Henkin wondered if other metals that are present in the body in minute amounts—zinc, for instance—played a similar role. Happily, the patients with idiopathic dysgeusia responded to zinc.

When Henkin examined Rudy, he found that Rudy's blood serum zinc and copper levels were both low and that Rudy's taste buds looked worn down, "moth-eaten." Two weeks after beginning zinc supplements, Rudy's taste buds looked normal and he could eat and drink normally again. To test the effect of the zinc, Henkin later gave Rudy placebos, and the disorder returned, much to Rudy's dismay.

Since treating Rudy, Henkin has become aware of and treated thousands of cases. One study has estimated that about 2 million Americans are suffering from hypogeusia or its related disorders but that only a tenth of them see a doctor about it. Some, like Rudy, develop the distortion after flu or a virus infection; some develop aberrations after surgery, perhaps as a result of medications, suggesting that some complaints about hospital food might be related to the patients' altered sense of taste. Other cases arise from trauma. Still others are idiopathic.

The zinc deficiency is the first nutrition-related cause to be identified. Henkin's group has found a zinc-containing protein in saliva, and they have determined that victims of hypogeusia often have a low level of salivary zinc, a more revealing indicator than blood zinc levels of the deficiency.

Space travelers also have a problem with taste. During the *Skylab* missions, the astronauts complained about the blandness of their meals. The second crew took horseradish and Tabasco sauce with them to jazz up the food, but the spices didn't help. Food in space is uninteresting, and it tastes different from how it tastes on Earth. Physiological factors seem to contribute to the lack of taste in space: Because the food volatiles may not travel their normal route to the olfactory mucosa in the low atmospheric pressure of the capsule, the sense of smell is not playing its usual part in taste. In the absence of gravity, body fluids tend to pool in the head and chest, causing congestion that may affect smell and taste. The Soviets tested their astronauts' taste buds aboard *Soyuz 30* and *31* with an electrical stimulating device, measuring dramatic changes in taste thresholds. S. Baranski, of the Military Institute of Aviation Medicine in Poland, suggested that during prolonged spaceflight, shifts in endocrine and metabolic functions may influence the performance of the taste buds.

Otherworldly Tasting, or the Fly's Feet

For animals, the chemosenses of taste and smell play key roles in detecting a suitable environment, finding food, and initiating reproductive behavior. Chemical perception was essential for survival in the murky aquatic environments that were the scene of so much evolutionary history; consequently, cells responsive to chemicals appeared early on.

Animals spend "money" in the form of energy and materials. It is cheaper to build a perfectly round cell than a cell covered with projections. Yet cells with hairlike or threadlike projections are found everywhere in the animal world, from protozoans to humans. The projections—called cilia, microvilli, flagellae—increase the surface area for detection of the chemical environment. They do dual duty because they also move. In one-celled animals, they can be used to move the whole animal toward or away from a chemical stimulus. Special proteins in the cell support the projections; these are the same proteins that are found in the muscle cells of higher animals. Sensation and movement are linked at the molecular level as well as at the behavioral and physiological levels.

Taste and smell—chemosensation—are difficult to separate in aquatic animals because of the medium. They are separated more easily in animals that live on land. Here, in our airy environment, we can talk about the sense of taste involving detecting chemicals in solution that directly contact the taste receptors and of smell as detecting airborne volatiles, vaporized chemicals.

We have tongues for taste. But are there other terrestrial designs for tasting? Well, animals that walk around in their food, for instance, might taste with their feet.

In the 1920s, D. E. Minnich of the University of Minnesota applied a sugar solution to the sensory hairs on the feet of butterflies. When the solution touched even a single hair, a butterfly uncoiled and extended its proboscis as it does when it feeds. Each hair seemed to be a taste receptor.

Later the biologist Vincent Dethier dipped the feet of blowflies into a variety of solutions. He observed which solutions caused extension of the proboscis and so was able to compile a long list of molecules that the fly found tasty. By 1950, largely as a consequence of Dethier's work, more was known about the taste receptors of blowflies than about those of any other organism.

Blowflies taste with 3,000 sensory hairs distributed among their six feet. They like sugar too. When a food-deprived fly steps into a sugar solution, its taste receptors send messages to the brain. The message is relayed to motor neurons that tell the proboscis to extend into the feeding position.

The fly proboscis is like a jointed soda straw with a sponge on the end, a kind of living mop. The bottom of the proboscis spreads open to expose a surface fringed with 250 hairs containing more gustatory receptors. When these contact the solution, they fire; in turn, motor commands initiate a sucking response.

The fly tasting experiments are interesting in their cleverness and also in the insights they give us about the fly design for life, as well as our own tasting mechanism. Each fly sensory hair has the same four receptors: one water-sensitive cell, one sugar-sensitive cell, and two salt-sensitive cells. The nerve fibers do not interact with each other; they carry their information, unaltered, directly to the fly's brain. The message is not as simple as just a yes or no, however. If a fly steps into a weak salt solution, it extends its proboscis. In response, a salt receptor fires 14 times each 100 milliseconds. If a strong salt solution is substituted for the weak one, the fly will suck only briefly and then retract its proboscis. Before this happens, the firing rate changes to 14 impulses each 70 milliseconds. The fly's

Blowflies and gourmets have essentially the same kind of taste organs—hairs. The hairs of blowflies just happen to be on their feet. Blowflies are 10 million times more sensitive to sugar solutions than humans are, and they ignore artificial sweeteners.

brain measures the frequency of firing of salt receptors and determines whether or not the solution is palatable.

Dethier originally measured the response of blowfly receptor cells to pure solutions. However, in nature, solutions—saps, nectars, plant and animal juices—are rarely just one thing or another; a blowfly steps into a lot of things, nourishing and nonnourishing. Dethier began to measure the blowfly's receptor responses to liquids with both sugars and salts, or salts and quinine, or sugars and amino acids. He found that salt added to a sugar solution depresses the activity of a sugar receptor, sugar added to a salt solution inhibits the activity of a salt receptor, and quinine inhibits the activity of all receptors. Some combinations of sugars in a liquid are especially excitatory and some are inhibitory.

Dethier suggested that a number of receptor sites on a sugar sensory neuron can bond with either sugar or salt molecules, others may bond with either sugar or quinine, and other combinations. If a solution contains both sugar and salt molecules, the molecules compete for the sugar-salt receptor sites. When salts or quinine occupy a site on a sugar receptor they do not cause the receptor to fire. The more sugar molecules in a solution, the more receptor sites will be occupied by sugars, resulting in larger potentials and a greater rate of cell firing.

Taste Aversions

Just as the natural aversion to bitter tastes can be overcome, taste aversions can also be acquired if the food or drink consumed induces illness. Laboratory experiments have shown, for instance, that if a rat is injected with apomorphine, it experiences a sudden, brief bout of nausea and illness. If it is given sweet food before the onset of the illness, it will develop a sweet-taste aversion, just as many people will show an aversion to a variety of foods consumed coincidentally before a bout of illness. I acquired an aversion to succotash in this way many years ago. (Some have suggested that for this reason sick children should be given junk food.)

Although this kind of conditioning and learning may make it more difficult to poison rats, it may be used in the opposite way with predators. A meal of hamburger laced with lithium chloride in capsule form causes nausea in coyotes and will make them reject hamburger next time it is presented to them (and hamburger on the hoof?). Similar taste aversions have been produced in rats, mice, cats, monkeys, ferrets, birds, fish, and reptiles, sometimes in as little as one learning trial.

At the Monell Chemical Senses Center in Philadelphia, scientists have identified a flavoring chemical, dimethyl anthranilate, that is readily accepted by mammals but unpalatable to birds. They suggest that it might be used to protect livestock feeds from consumption by birds. They have also identified a sunflower hybrid whose flavor (from anthocyanins in the seed hulls) is strongly repellent to birds but not to mammals. It has a potentially high oil yield, and they expect farmers in the High Plains, where bird damage is most severe, to be interested in it.

Keeping up with Taste

In 1499, at the end of a two-year voyage to Asia, Vasco da Gama brought back spices to Portugal that were worth 60 times the cost of his round trip. Spices had been valuable for centuries as seasonings, and because they covered the taste of tainted food, they gave a starving and crowded Europe access to more calories, perhaps their greatest value. During the Middle Ages, pepper was counted out in individual peppercorns, and a pound of ginger was worth a sheep. For three centuries, western Europe fought wars for control of the spice trade.

Human diets have changed over the centuries. Humans are the consummate omnivores, able to meet nutritional and hedonistic requirements from an incredible variety of foods. (Think of the koala bears that must eat eucalyptus leaves; the pandas that eat bamboo; the hawk that consumes only snails—how precarious their existence, and how incumbent upon humans to preserve it.) We are descended from nibblers, and our distressing modern tendency to get fat is related to the civilized habit of eating regularly scheduled meals instead of just obeying the brain's command to eat small amounts of food frequently. We do both.

The part of the world we live in makes a difference in what we learn to eat: Polynesians eat fish, breadfruit, bananas, coconuts, the potato-like taro; Hindus eat rice, wheat bread, chickpeas, mangoes, dairy products; Guatemalans eat beans, corn tortillas, chard, cheese, chicken; the Japanese eat rice, mandarin oranges, fish, soybeans, seaweed, green onions; the French eat bread, potatoes, artichokes, chicken, cheese; Americans eat bread, milk, potatoes, apples, corn, beef. Some diets, perhaps the strangest, are mostly animal. The diets of Eskimo (until the supermarket and television came to the cold North) have been high in animal fat. The cattle-herding Masai of Africa eat meat, milk, cattle blood (drained carefully from a living animal, to conserve the resource—like parasitism), and a little fruit for vitamin C.

I grew up in a rural area of the United States eating a lot of bread and bacon gravy. At that time, sour cream was not something you would want to put on a potato (but it could be used to make cookies), and wine was drunk by winos. We ate more eggs and cheese than meat, and canned peaches, pears, and applesauce must have warded off winter scurvy. I saw my first avocado when I was a teenager. I saw and savored my first kiwi fruit in 1970. Jicama, squid (Mexico), and snails (anything tastes good in garlic butter) came later.

Today I have an incredible array of choices before me when I go to the supermarket. Refrigeration and shipping have made certain foods available all year. Some exotic fruits and vegetables are being grown closer to home. New processed foods are constantly introduced, based on scientific study and market research.

Availability has changed since da Gama's time. But we are always looking for novel tastes to stimulate eclectic palates. I now see hairy kiwi fruits in abundance and at low prices. I could go to the grocery store and get one right now, in ten minutes. But maybe I'd rather have a hot fudge sundae.

For an animal is by our definition something that has sensibility, and chief of all that has the primary sensibility which is that of Touch, and it is the flesh which is the organ of this sense.

—Aristotle

See! how she leans her cheek upon her hand:
O! that I were a glove upon that hand,
That I might touch that cheek.

—William Shakespeare

Pale hands I loved beside the Shalimar,
Where are you now? Who lies beneath your spell?

—Laurence Hope (Mrs. M. H. Nicolson)

. . . the screws of his skulled head
joining the screws of his hands,
pink convolutions transmitting to white knuckles
waves, signals, buttons, sparks

—Carl Sandburg

What is it, then, this seamless body-stocking, some two yards square, this our casing, our facade, that flushes, pales, perspires, glistens, glows, furrows, tingles, crawls, itches, pleasures, and pains us all our days, at once keeper of the organs within, and sensitive probe, adventurer into the world outside?

—Richard Selzer

Hug me, you little drumkin.

—My mother

S · I · X

▼

Touch: The Confirmatory Sense

TOUCH (derived through Fr. *toucher* from a common Teutonic and Indo-Germanic root, cf. "tug," "tuck," O. H. Ger. *zucchen*, to twitch or draw), in physiology, a sense of pressure, referred usually to the surface of the body. Sensations of temperature and those peculiar sensations called pain are also referred to the skin.

Sensory Enfoldment

Touch gathers information about the world, confirming the information of the other senses, confirming what we hear, see, or smell. Touch asks, what is it? where is it? It is the haptic sense, the somesthetic sense, giving us, via experience, an extended sense of living and acting in space. It puts us in imaginative physical contact with objects once touched.

It is a magic transformer, reaching out to the world and to others. In the Sistine Chapel, God reaches an index finger toward Adam's lifeless hand to transmit the spark of life. Healers use the laying on of hands. An extension of the hand touches too. Magicians and fairy godmothers use wands, extensions of their hands, to work their changes. By a cane, a blind person confirms. By a racket or bat, an athlete responds. My pencil touches the paper; I feel it. Touch is both active and passive: I can concentrate on the edge of the table or on the dent the table makes in me.

Babies are touched—nuzzled, carried, attended to, enfolded. Babies are also the great touchers, explorers, learning, using sensitive fingers, tongues, and lips to confirm what they see. They explore, then run back to adults for the tension-reducing touch. The touching of sexual intimacy carries its own kind of rhythm and natural progression. Lifelong, we need embrace, we need the exploration of touch. We run our hands over the objects in our environment, and others touch us. We grasp hands. We hug. We pat. We stroke. We straighten a collar or a tie, pick lint. We go to the beauty shop, the barbershop. We touch. We feel. We are touched. Touché.

Via the skin. The skin may be considered the largest organ of the body, and the largest sensory organ—the organ of cutaneous sense. The total area of skin surface of an average-size adult human is about 1.8 square meters, about 1,000 times the area of the eyes' retinas. The skin and the nervous system develop from the same layer of embryonic tissue. As development continues, the nervous system folds inward, zips up, and the skin remains as a kind of external nervous system, its different sensory structures picking up information about pressure, heat, cold, and pain. The covering becomes functional early in the womb; a fetus responds to touch at about nine weeks.

Upon close examination with a low-power lens, the apparent smooth surface of the skin becomes a speckled landscape of ridges and valleys and pores. With higher power, we see the skin as an ecosystem, supporting microscopic flora and fauna in diverse ecological niches: the desert of the forearm, the woods of the scalp, the tropical forest of the armpit, the damp basin of the navel, the deep narrow well of the hair follicle. Bathe as we might, we are not alone.

The skin of ruddy lips, nipples, and genitalia is highly vascularized, rich in blood vessels. The skin of the palms and soles is thick, ridged, and hairless—glabrous; the remainder of the skin is thinner and hairy to different degrees. The scalp, for instance, possesses about 100,000 hairs.

The *epidermis* consists of an outermost layer of tough and flat dead cells, and the germinative layer below that replaces shed cells. Below the epidermis, the *dermis* is penetrated by various nerve endings and tiny capillary loops. Farther down are the larger vessels, sweat and sebaceous glands, hair follicles, fatty and connective tissue. The coiled *sweat glands* (the "fairy intestines" of Oliver Wendell Holmes) bring water to the surface, cooling the body, including the skin, through evaporation.

The skin provides us with the primary sensations of touch, pressure, temperature, and pain. The other sensations, such as tickling, itching, burning, creeping, and crawling seem to be derived from combinations of these. Certain perceptions of tactile stimuli release specific defensive reactions. For instance, creeping things on the

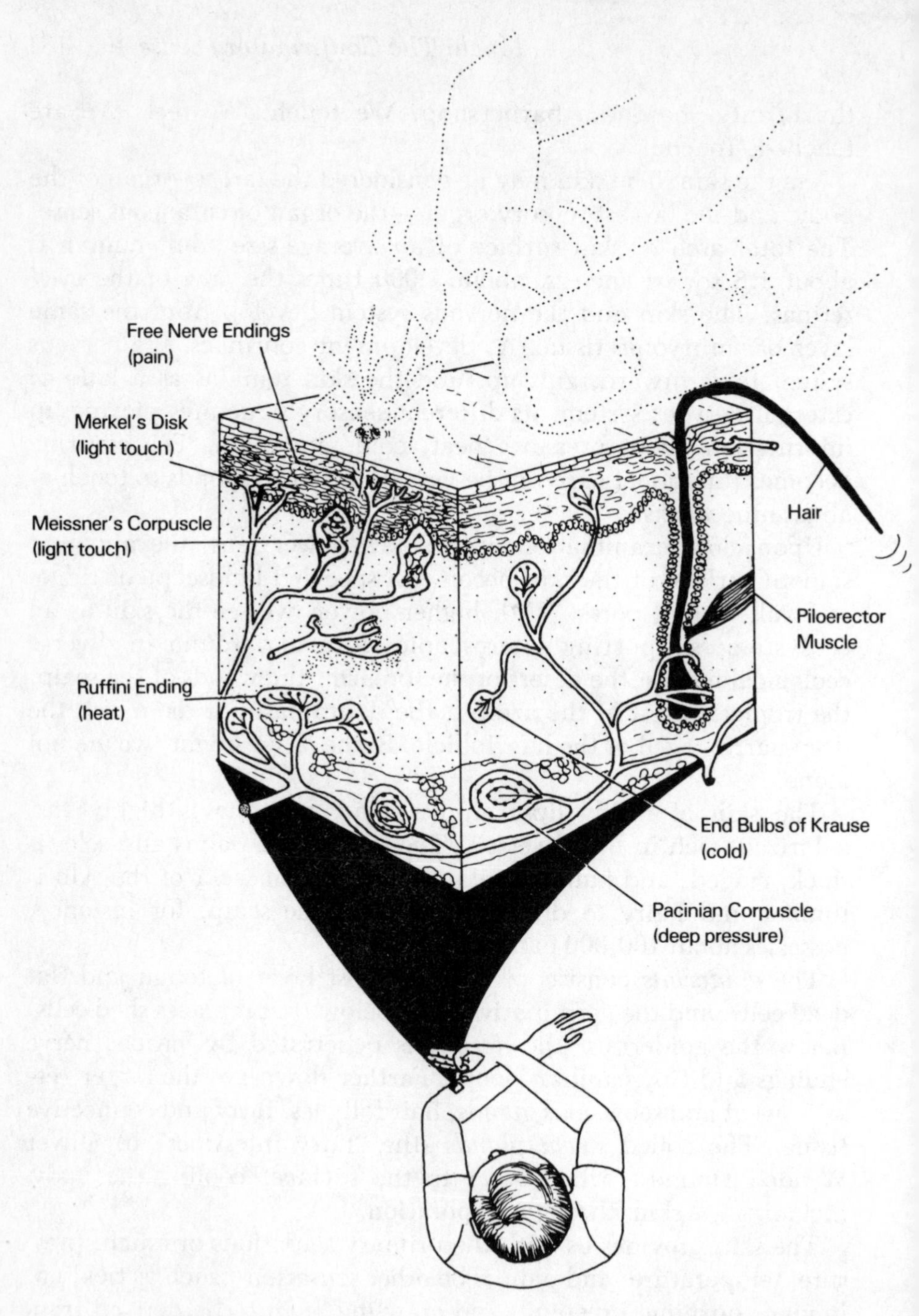

The skin, a versatile and complicated wrapper, is our first line of defense against invading organisms. It is also used in temperature regulation, excretion, and perception of a variety of stimuli.

back of the hand release a shaking movement, which Konrad Lorenz interprets as a defense against insects.

The names of the skin receptors provide a nice example of the old days of physiology, when every newly described structure was named after someone. The egocentric practice has conferred a sense of history and people to the study of anatomy—who were these guys looking at these little things anyway?—but today the trend is toward naming structures in terms descriptive of their functions. My ophthalmologist speaks of a "trabecular meshwork" in the eye. "The canals of Schlemm?" I ask. "Yes, yes," he says.

So. The best described and best understood (most recognizable) receptor is the *Pacinian corpuscle*, a nerve ending encapsulated like an onion and moderately distant from the skin's surface. A deep pressure, tension, and vibration receptor, it dwarfs all other skin receptors and seems to have a more private communication with the nervous system. The palm side of the hands and fingers are rich in Pacinian corpuscules, with 1,000 to 1,500 of them, their density greatest in the finger pads and distal joints. They are also found near blood vessels and lymph nodes, other joints, the mammary glands, and genitalia. Deep pressure sensation is felt over a large area, is long lasting, and has little variation in intensity.

The other receptors are more equivocal in structure and function. M. von Frey's 1890s idea—that Meissner corpuscles, Krause end bulbs, Ruffini cylinders, and free nerve endings are responsible for the different qualities of touch, cold, warmth, and pain, respectively—has persisted for a long time, but recent evidence indicates a more complicated and integrated role of cutaneous receptors. One text lists 13 receptor types identified from a cat's skin with the stimuli they seem to best respond to, such as steady displacement of a hair, rapid displacement of a hair, steady indentation of skin, or rapid indentation of skin. Clearly, the skin is sending complicated messages.

The skin receptors, understandably, have presented some problems in study. Some researchers have localized sensations on their own skin, then excised the bit for examination under the microscope. An experiment reported from the 1930s involved the determination of warm and cold spots on the prepuce from both sides of the skin fold. This experiment received some notoriety because of the self-discipline required of the experimental subjects, who had their foreskins suspended from "dulled" fishhooks during the course of the experiment.

The receptors known as *Meissner's corpuscles* are located in papillae or little islands of the skin and consist of a mass of nerve cell processes known as *dendrites* enclosed in connective tissue. *Merkel's discs* are flattened formations of dendrites attached to deeper layers of epidermal cells. Both impart the sense of light touch when they are deformed, and they are found in the eyelids, tip of tongue, lips, nipples, genitalia, fingertips, palms, and soles. The skin of a three-year-old has about 80 Meissner's corpuscles per square millimeter, a young adult 20, and the elderly about 4. Root hair plexuses are dendrites associated with hair follicles, so that mechanical displacement of a hair is also felt as light touch. The *end bulbs of Krause* are oval connective tissue containing dendrites. The *end organs of Ruffini* are less numerous and are deeply embedded in the dermis.

Pain receptors are least differentiated and found as bare nerve endings, widely distributed throughout the skin and mucous membranes of the body.

Estimates are that the skin has 50 receptors per 100 square millimeters, a total of 640,000 sensory receptors. Tactile points vary from 7 to 135 per square centimeter. More than half a million sensory fibers enter the spinal cord, capable of taking messages to the brain.

The distribution of receptors varies, however. If a hairpin, spread so the tips are about an inch apart, is run lightly down the skin of the forearm, past the wrist and down the skin of the palm, it feels like a single touch that mysteriously branches into two separate touches on the palm. With a compass, the spatial thresholds of receptors can be measured in different parts of the body.

A mapping exercise, similar to the tongue-mapping exercise, also gives students a picture of their skin senses. With a washable pen, they draw two-inch grids, ruled into quarter-inch squares, on the back of their hand, palm, and inside forearm. They draw three similar grids on paper. Then they touch each square of skin with the flat head of a pin, lightly, then firmly; the sharp point of the pin; a paper clip tip chilled on ice; and a paper clip tip heated in water. The sensations—light touch, deep pressure, pain, cold, and heat—are recorded with different colors, and a sense map results.

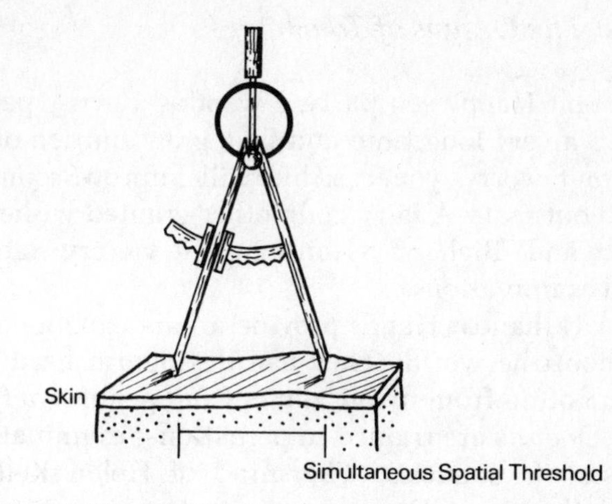

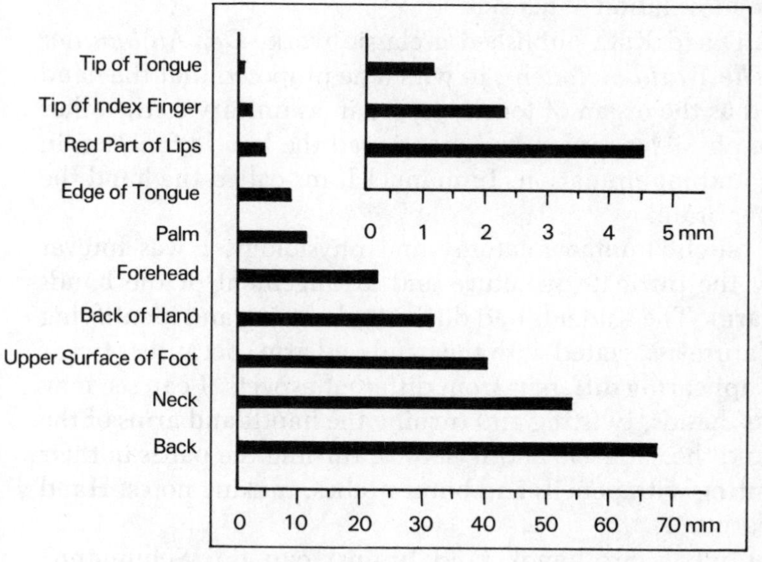

The differential population of receptors on the skin is demonstrated by a two-point discrimination or spatial threshold test, using a compass. The compass is applied to the skin, and a person indicates when two points are felt and when only one is felt. The more sensitive the area, the closer the compass points may be placed and still felt separately. The tip of the tongue feels two points only 1.2 mm apart. On a person's back, receptors are fewer and farther apart. (From Fundamentals of Sensory Physiology, *edited by Robert F. Schmidt, © 1981. Redrawn with permission from Springer-Verlag, New York, Inc. and Dr. Robert F. Schmidt.)*

Hands: The Organs of Touch?

At a recent Halloween party, I went as a gypsy palm reader ("You will live a very long time and be a great burden on your children. And don't worry, your sex life will improve soon."). A mailman handed out mail. A fairy godmother granted wishes with the touch of her wand. Richard Nixon gave the victory sign. Three doctors offered examinations.

All with hands. Hands provide an amazing amount of information about the world. They can distinguish hard metal from soft rubber, cotton from nylon, dimes from pennies, a fever from a cold sweat. Doctors are trained in palpation—examination by touch for purposes of diagnosis. The mind of Helen Keller was created through the stimulation of her skin.

In 1925, David Katz published a classic work, *Der Aufbau der Tastwelt* (*The World of Touch*), in which he proposed that the hand be regarded as the organ of touch, an organ as unitary as the other senses. The physician Galen had mentioned the hand's function in perception and manipulation. Immanuel Kant called the hand the human outer brain.

When I taught human anatomy and physiology, I was forever amazed by the intricate structure and arrangement of the hands and lower arm. The students had difficulty learning and identifying the musculature associated with the hand and arm, because it twists and turns, appearing different from different aspects. I can see now the students' hands, twisting and turning the hands and arms of the manikins and the cadavers and skeletons, turning the pages in their books, pointing with pencils and blunt probes, making notes. Hand in hand, learning.

Consider what our hands (and brains) can do: Schumann's C Major Toccata, opus 7, has 6,266 notes and requires 4 minutes and 20 seconds to play. A speed of 20 to 30 notes per second requires 400 to 600 separate actions of just the muscles of the fingers.

And consider what our hands tell us: Close your eyes, spread the fingers apart, then draw them over a surface, and you will experience the surface as extending into the paths between the fingers, projecting the object into space. Our brains integrate and fill in the empty space, much as we fill in the information for the blind spot of the eye. We hold a ball, for instance, and information from ten different digits tells us that it is a single object. The sensitive fingers provide us with memory images. I know what the items in my house, in my office, on the way to work, feel like because I have touched them. I confess to touching a statue in a museum above a sign that said, "Do not touch." It was a compulsion. I saw that

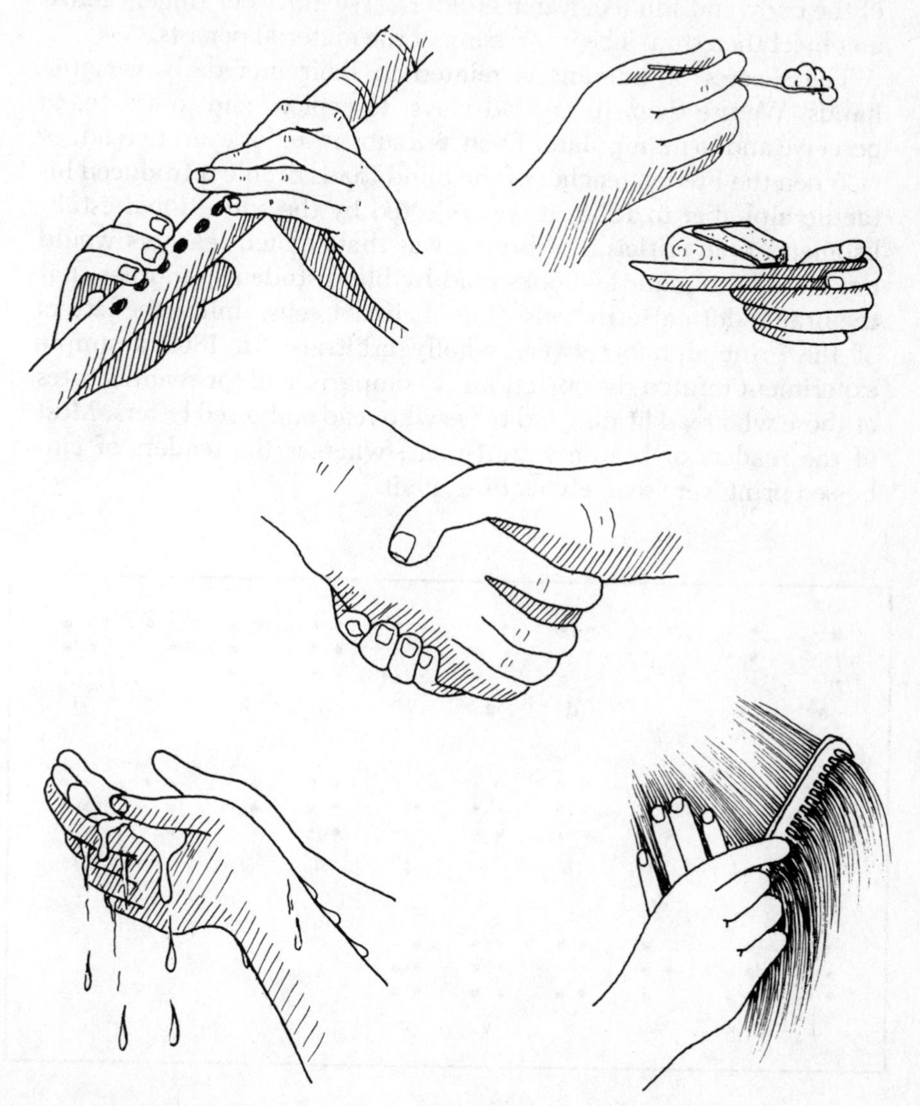

The human hands with their opposable thumbs can perform a seemingly endless variety of small-motor athletic activities.

others had touched it, too, and I empathized with children who are told not to touch.

Fingertip skin is thick and resistant to shifting; at the same time it is cushiony, adapting its shape to that which is touched. Fingertips provide afterimages. We can move an object to a less sensitive part of the body and still feel it in its fullness. We move our fingers across an object then stop. The impression of the material persists.

The success of humans is related to their incredibly versatile hands. We use them in myriad ways: to explore and to create, to perceive and to manipulate. Even as a substitute for eyes, to read.

When the French teacher of the blind Louis Braille introduced his tactile alphabet in 1852, it was rejected by the educational establishment. One criticism of braille was that sighted teachers would not be able to read the books read by blind students; another that the braille dot patterns consisting of six-dot cells, unlike the letters of the print alphabet, were wholly arbitrary. In 1869 a simple experiment refuted the objections: a comparison of the reading rates of those who read braille and those who read embossed letters. Most of the readers of braille were fluent, whereas the readers of embossed print were scarcely reading at all.

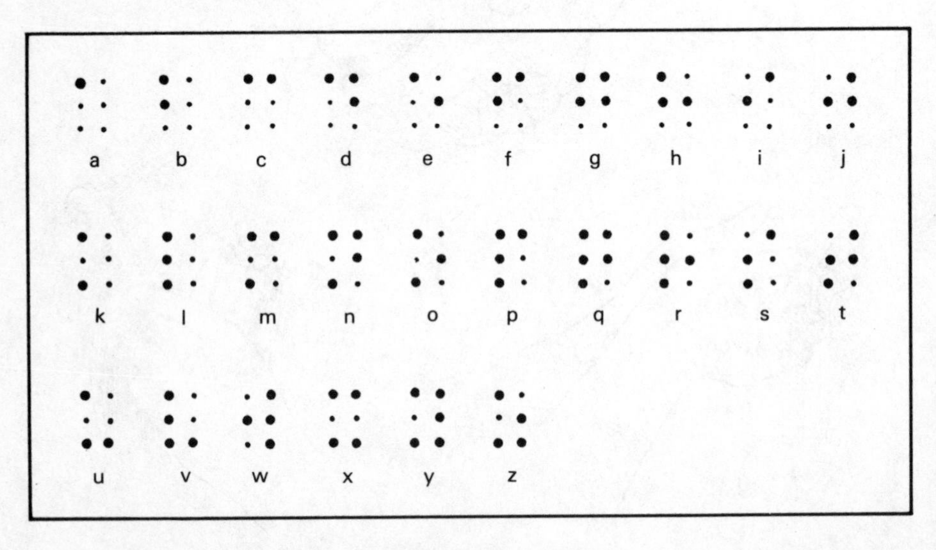

The embossed braille alphabet brings written language, thus more information and communication, to blind people.

Although braille can be read much faster by touch than embossed print, the process is much slower than the visual reading of print. The skin and the fingers seem simply not to permit the same rapidity of reading processing that occurs with vision, although both systems ultimately yield information that is mapped into language. On the average, blind junior high students read braille at 60 words per minute, senior high students at 80 words per minute, and adults at 104 words per minute. For comparison, the average silent visual reading rate for high school students is between 250 and 300 words per minute. A very few braille readers can read that fast. Those who use two index fingers rather than one have been found to read faster. Apparently, the trailing index finger clarifies and evaluates the information acquired by the leading index finger. Braille has been adapted for writing various languages and for transcribing music, mathematics, and scientific symbols.

Temperature Monitors

Humans are originally tropical creatures, intolerant of skin temperatures much below 30 degrees Centigrade. We're warm-blooded, or homeothermic, organisms, maintaining a constant body temperature. We function best within a certain narrow temperature range. We tend to avoid high temperatures that can cook us and low temperatures that can freeze us. When the environmental temperature is hot, our blood vessels dilate and we perspire, both of which help us to lose body heat. Cold, we shut down peripheral circulation, shunting the blood to our more vital internal organs. To exploit cold areas of the world, we have invented elaborate cultural adaptations to help us maintain our skin temperature. (Humans have some physiological adaptations as well: Eskimo are generally stocky with short appendages, exposing less skin surface to the world than Watusis of the same weight.)

Our heat and cold receptors monitor the environmental temperature. They are distributed over the skin in varying densities, less numerous than touch receptors. Cold receptors lie just below the epidermis, and the warm receptors lower, in the middle of the dermis. We have more than twice to about 12 times as many cold points as hot points overall. Their greatest density is in the most temperature-sensitive region of the skin, the face. On the face are 16 to 19 cold points per square centimeter; on the face, warmth sensitivity can't be resolved into individual points—it forms a continuum.

I like to read in the bathtub. When I first get into a warm bath, at about 33 degrees Centigrade, it feels warm. This pleasant sensation fades away soon, and I've become adept at turning on the hot water tap with my toes (so I don't get the book wet). The water hasn't cooled that much; rather, my skin's heat receptors have adapted. I've also felt the opposite phenomenon, jumping into a pool at about 28 degrees Centigrade on a hot summer day. The water at first feels cool, but after a while the sensation of cold dies away.

In an intermediate temperature range, warming or cooling produce only temporary sensations of warm or cold; adaptation to the new skin temperature is essentially complete. This range is the neutral, or comfort, zone. Above or below the comfort zone, permanent sensations of heat or cold are produced even when the skin temperature is kept constant for a long time. Partial adaptations take place when hands are immersed in hot dishwater, for instance. Immersion at 42 degrees Centigrade initially results in a strong sensation of warmth. This falls off rapidly at first, then more slowly, until a maintained sensation of warmth at low intensity is reached. The same is true of cold. For instance, sometimes my feet feel cold for hours on end, and it seems to take forever to warm them. The upper and lower limits of the comfort zone vary somewhat; the limits expand for a smaller area of skin. However, in experiments with naked humans in a climate-controlled room, the comfort range narrows to 33 to 35 degrees Centigrade.

Persistent sensations of warmth are intense. At temperatures of more than 43 to 44 degrees Centigrade, the sensation of warmth gives way to painful heat sensation. At temperatures below 30 degrees Centigrade, the maintained cold sensation increases in intensity, the colder the skin. Cold pain sets in at skin temperatures of 17 degrees Centigrade and lower, but at skin temperatures as high at 25 degrees Centigrade the sensation of cold is unpleasant, especially when large areas of skin are affected. If cooling takes place very slowly, a person may not notice that large regions of the skin have become quite cold, with concomitant loss of heat from the body, especially if one's attention is distracted by other things.

After-sensations are common for the sense of temperature. If you press a cold metal rod against your forehead for 30 seconds and remove it, a sense of cold persists even though the skin is warming up again. Recordings have shown that the greatly cooled cold receptors continue to discharge during rewarming, at first even at an increasing rate. Very strong warm stimuli, such as too-hot bathwater, will produce a paradoxical cold sensation. This is probably because the cool receptors, normally silent above 40 degrees Centigrade, discharge again transiently when rapidly warmed to more than 45 degrees Centigrade.

My biology students liked E. H. Weber's simple "three-bowl" experiment. We filled one bowl with hot water, one with lukewarm water, and one with ice water. One hand (or the index finger) is immersed in the hot water, the other in the cold water. Then both hands or fingers are moved to the bowl of lukewarm water. The hand that has been in the hot water will get a sensation of cold, and the hand that has been in the cold water a sensation of warmth.

Although the skin of the hands is extremely sensitive to touch, it is less sensitive to temperature than are other parts of the body. Hands will feel comfortable without gloves on a chilly day. When my sister warmed her babies' bottles, she tested a drop on her wrist rather than her finger; the wrist is a better sensor of body temperature. Likewise, lips touched to another's forehead are a better sensor of fever than are the hands.

If you touch an ice cube or the inside of the freezer (don't do this with your tongue), it seems "burning hot," because the skin has receptors for sensations of cold and warmth but none for very hot temperatures. Very hot temperatures stimulate a combination of cold and pain senses, producing a warning signal to prevent a dangerous burn. Very cold temperatures also stimulate the pain receptors and are easily confused with the hot sensation.

Pain: Hurtful, Warning

Pain may be caused by special pain receptors in the skin, usually naked nerve endings, or by overstimulation of pressure and temperature receptors. Pain sensations vary: People report pricking, burning, itching, and aching (and more creative sensations too.

One doctor collected patients' descriptions of pain, which included "it feels like antelope jumping on my bones"). The sensation felt seems to depend on how strongly the nerve endings are stimulated.

Unlike the other sensory receptors, pain receptors adapt only slightly or not at all, which makes biological sense; if animals adapted to pain, it would cease to be sensed and irreparable damage could result.

Newborns do feel pain, but this is a recent recognition by the medical establishment. Surgeons have often operated on babies without the benefit of anesthetics, assuming that babies did not feel pain and that potent anesthetics posed a threat to them. However, anesthetics have become safer, and mounting evidence about the newborn's capacity for pain has resulted in recent policy statements from various groups urging the use of painkillers when newborns must have surgery.

Sensitivity to pain varies throughout the body, depending on how many receptors are present. The distribution of pain receptors is different from the distribution of touch and temperature receptors. Pain receptors are clustered thickly on the surface of the eye, so that an almost-invisible piece of lint feels like a jute rope. We blink rapidly to get rid of it. On our fingertips are a large number of touch receptors, but only a few pain receptors, which would otherwise interfere with our sense of touch. The amount of pressure needed to cause pain in the fingertip is about 1,500 times as great as that which will cause pain to the eye and three times as great as that which will cause pain on the back of the hand.

Pain is a useful warning to possible damage to the body. Pain has no projection outside the body, but it has undoubtedly been of great survival value in evolution in providing warning of damage or destruction of the body. The experience of pain itself is not enough. Survival is related to the reactions, fight or flight—to which pain gives rise. We learn to link the senses with events in space outside the body, substituting pain symbols, such as fire, for bodily pain, so that defensive reactions begin when the stimulus is still just a threat. We avoid potential painful situations or objects.

Extreme pain is usually accompanied by immobility. For an animal in pain to remain immobile may seem maladaptive, but a major function of pain may also be to promote healing. The body may produce its own opiate chemicals in response to pain, so that an injured person or animal is kept still and injured areas are healed.

However, the circumstances people find themselves in have an effect on the amount of pain that is perceived. Boxers and others are conditioned to withstand pain. Stories are apocryphal about injured soldiers, athletes, or accident victims seeming to ignore their pain until some vital task is performed. In these cases, the pain signals may be intercepted on their way to the brain by competing signals.

We know that pain can be relieved by inhibiting it by other sensations such as touch and pressure, so we scratch, rub, and press a painful area. The same principle works for visceral pain. A blistering mustard plaster or a heating pad will stimulate the nerves of the skin, inhibiting the nerves of the corresponding abdominal region. Thankfully, I escaped mustard plasters, but I can attest that the heating pad treatment works for menstrual cramps.

Acupuncture is also helpful for pain, seemingly for the same reason: Transcutaneous nerve stimulation puts noise into the system, and the brain has difficulty interpreting the competing stimuli that it is receiving. The gate-control theory proposes that because of the size of the nerve fibers, the impulse passing along the touch-conducting neurons travels faster than the pain-conducting neurons and thus has priority in the brain, closing the gate to pain. The effects can last for several hours. Acupuncture is currently used in the United States mostly for childbirth, tic douloureux (trigeminal nerve inflammation), arthritis, and other nonsurgical conditions.

One of medicine's oldest and most fundamental goals has been the conquest of pain and the search for painkillers. The body's main messenger of pain is a chemical known as bradykinin, which is created when body tissue is injured. When a cell is injured, enzymes are activated that clip the bradykinin free of large precursor molecules. The bradykinin then works by fitting into a precisely shaped receptor on a nerve cell, causing the nerve cell to fire and notifying the brain that injury has occurred. It also opens leaks in small blood vessels near the injury so that the area is bathed in fluid and infection-fighting white blood cells. It sets off the production of chemicals that promote inflammation and boost the pain impulse further. Aspirin (which we've used for 100 years without knowing its secrets) and related painkillers work by blocking one group of these chemicals, the prostaglandins. Bradykinin is broken down quickly, but it also stimulates the production of more bradykinin until the injury is under control. Several bradykinin blockers have been synthesized, and research on their effectiveness is continuing. A Brazilian snake, *Bothrops jararaca*, has been useful in bradykinin research; the snake's venom causes the liberation of bradykinin and also inhibits its breakdown. Its bite is extremely painful and causes a drastic drop in blood pressure.

For awareness of pain to occur, the message must get through to the interpretive centers of the brain. Pain may be terminated by anesthesia, sleep, or hypnosis; in those cases, the reticular activating system of the brain stem has shut down and messages are not being transmitted.

Pain is extremely complicated and individual, with both physiological and psychological aspects that may be difficult or impossible to tease apart. A psychologist who specializes in treating pain tells me that he often sees a relationship between stress, the quality of life, and how pain is experienced. In general, the more people are interested in their world and the more they like their lives, the less they focus on their own pain. In some cases, pain becomes a way of relating to the world and to others, and an entire family must learn new ways of dealing with each other.

He tells me of treating a woman who had had headaches for 25 years. She had lived on an isolated Western ranch. Two days after checking into the pain clinic, with no treatment except the companionship of others in the hospital, her headaches disappeared. She was lonely and didn't know it or admit it. Headaches can be a good barometer of the quality of life, he says. General body muscle pain is a frequent problem too. The fight or flight response was adaptive in our evolutionary past, yet in modern society the solution to most problems is not to smack someone or to run. We can't discharge the response, and so it manifests itself in chronic muscle tautness and pain.

Chronic pain is an expensive health care problem, yet training of physicians has not focused much on pain management, the psychologist tells me. Perpetuation of pain may develop as a kind of reflex action at some point in the nervous system. The medications that are prescribed can be part of the problem, and vicious cycles can develop. In some cases, he says, the body seems to produce pain to get the painkiller. Acute pain is generally a symptom of disease; chronic pain becomes in itself the disease.

If we can see an injury, it is easy for us to visualize the pain as coming from the injury. We can locate the origin of the pain based on our past experience and accurately project it back to the stimulated area. In the case of internal pain, however, the pain may be "referred" to the skin over the stimulated organ or somewhere distant. The area to which the pain is referred generally receives innervation from the same segment of the spinal cord, so that the pain of a heart attack is felt in the skin over the heart and along the left arm. Liver and gallbladder pain may be felt on the right side of the neck, kidney pain in a girdle that extends over the hips and into the groin. In phantom pain, a person feels pain from an amputated limb as a result of stimulation of the peripheral nervous system and the central nervous system's interpretation of the signals as coming from the nonexistent limb.

In the case of sunburn pain, the threshold of sensitivity of nerve endings is lowered, and the sunburned skin is tender to all forms of stimulation.

About 5 percent of the population is not very sensitive to pain, and in this hyposensitive group are some who are unable, because of nerve damage, brain tumors, or a genetic disorder, to feel pain at all. A psychologist tells me of patients who inflict injury upon themselves in order to "feel." Some children have been identified who can sustain fractures and gross burns without being aware of them, which is obviously not adaptive. And if the children cannot feel pain, how might they be taught not to hurt others?

The sense of touch can also have its dysfunctions. Clumsy children, for instance, children who are prone to break things and regularly spill milk at the table, may have difficulty interpreting information that comes in via touch. The disorder, known as *sensory integrative dysfunction*, was described several decades ago but has been gaining wider recognition in recent years. Some physicians dismiss the disorder as something a child will outgrow, and it does seem to become less apparent with age. However, affected children may dislike being touched by others, finding the sensation of light touch unpleasant and disturbing.

Social Contact

Tactile communication is important throughout the animals and, as one might suppose, especially in animals in which fertilization is internal and so must get close to mate. Courtship behavior breaks down space barriers and reduces aggressive behavior (think of a black widow male mesmerizing a female by twanging her web), as well as saying "I'm attractive, ready, and fit. Mate with me."

Touch plays an especially large role in the lives of primates. We and our relatives are social, contact animals. Primate mothers carry their young on their bodies for long periods, and the adults often sit or sleep close together. Primates lick and groom one another, cleaning each other and removing parasites. In baboons, tactile behavior progresses from an infant's suckling to grasping and sucking fur to the adult pattern of parting fur and removing fine particles with its lips and fingers.

Some have suggested that grooming behavior has played a role in development of fine coordination of the primate hand. Grooming is also social cement, and patterns vary among groups. Primates may nuzzle, embrace, tickle, and bite each other. Chimpanzees pat each other's hands, faces, and groins. They lay hands on each other's backs in reassurance and kiss in affection.

Our language is rich in reference to touch and its emotional links. In recent years researchers have begun to study nonverbal communication—body language—and the role of touch in a variety of contexts. Touch plays critical roles in human social relationships, including infant bonding and attachment, caretaking, aggression, courtship and sexual relationships, first impressions, children's interactions, and in psychotherapeutic and medical settings.

It was formerly thought that infant humans and other mammals formed attachment bonds with their mothers via suckling and feeding. In the 1950s, however, Harry F. Harlow reported on his studies with infant rhesus monkeys raised with surrogate wire-frame mothers. The infants did not bond with the surrogate that provided milk; they bonded with the terrycloth surrogate that provided contact comfort. Further studies showed that infant rhesus monkeys deprived of contact with their mothers did not form appropriate social bonds and were not good mothers themselves, although they were reproductively normal. Harlow speculated about the existence

Humans go to the beauty shop or the barbershop. Baboons go to other baboons. Grooming keeps ectoparasites at bay, helps manage tension, and is an important factor in social solidarity.

of a "critical period" in which physical contact and stimulation were essential for the development of proper social behavior.

These kinds of studies have been greatly expanded, and in the last two decades primatologists have learned much more about the fundamental role of touch in social attachment. Attachment bonds are critical to normal development and functioning of primates, especially humans. The presence of attachment bonds is associated with pleasure and good health; their disruption is associated with discomfort, and inappropriate disruption may have negative effects on the animal, both psychologically and biologically. The belief in the healing powers of touch may relate to its role in the formation of attachment bonds and the associated feelings of pleasure and good health.

Desmond Morris has proposed a 12-step sequence of physical intimacy shown in human courtship, a kind of flow chart of body accessibility. Each stage depends upon the couple's passing through the previous stages. As the pair moves toward greater physical intimacy, each assumes that certain touches are acceptable and expected. If a touch is not reciprocated, that also sends a message. The reciprocity of touch is important in relationships between parents and children and between friends.

Touch and Development

The senses need to be stimulated to develop, and many studies indicate that humans and other mammals need certain kinds of tactile experiences to develop into healthy individuals. Newborn mammals seem to need their mother's licking. Licking can orient them toward her nipples for sucking, and licking by their mother after they have fed on her milk provides a critical stimulus to the gastrointestinal tract and the genitourinary tract—the mother is essentially teaching the young to urinate and defecate.

Studies with rats have shown that handled offspring have better immune systems and gain weight better. Early handling also leads to a permanent sensitivity of the part of the brain that controls the stress response, resulting in a reduced level of stress hormones that can damage brain cells. Gentled rats are less disturbed by stress and starvation; they learn best and survive longest. This fact has posed some problems in research. Scientists may get different results with gentled rats than with rats that are stressed.

The stimulation between mother and offspring works both ways. Stimulation of the mother by the suckling young and her self-licking stimulate the production of the hormone prolactin, which helps the mother recover from the trauma of birth and maintains milk production or "broodiness." The attachment bond is adaptive for both offspring and parent. The parent helps the offspring survive and grow up to reproduce. The offspring perpetuates the parent's genes for the behavior.

Of course, human mothers do not lick their young. But Ashley Montagu has suggested that the uterine contractions during labor may serve the same function, providing massive stimulation of the fetal skin. The average duration of labor for a firstborn human is 16 hours and for subsequent children 8 hours.

Montagu also writes that caesarean-born children are not as responsive as children delivered vaginally, and notes that breastfeeding has beneficial effects for both mother and child. Suckling accelerates contraction of the mother's uterine muscles; the uterine vessels contract, the uterus reduces in size, and the placenta is detached and ejected. A mother holds the suckling child at alternate breasts, giving equal stimulation and exercise to both sides of the infant's head. Breastfed children are superior in speech development, Montagu writes, and this is probably related to the exercise of the right muscles and to the mother talking to the child as she feeds it.

Without touch, children die. As late as the second decade of this century, the death rate in U.S. foundling institutions for infants under one year old was nearly 100 percent. Dr. Fritz Talbot of Boston brought the idea of tender loving care back with him from Germany in the 1940s; he had visited a children's clinic in Dusseldorf and saw a fat woman carrying a puny baby on her hip. The clinic director told him that when they had done everything medically for a baby and it still didn't do well, they turned it over to Old Anna, who was always successful. Introduction of "mothering" regimens in institutions dramatically reduced the mortality rates of children.

Recent studies have shown that premature infants who were massaged for 15 minutes three times a day gained weight 45 percent faster than others who were left alone in their incubators. The massaged infants did not eat more than the others. Their weight gain seemed to be related to effects of touch on their metabolism.

Physical contact is the ultimate signal to infants and to small children that they are safe. Children, say psychologists, have many ways of telling parents they want touch. They may cling and demand help with things they've been doing perfectly well themselves, ways of seeking reassurance when they are learning to do many things. At about age one, children frequently develop an attachment for a soft cuddly blanket, pillow, or toy, dragging it around with them, stroking and caressing it. It's there, providing the security of touch, when the parent isn't. At about 18 months of age, toddlers may exhibit separation anxiety and insist on parents' company, especially at bedtime.

Touch separates "me" from "not me." It is important in forming a body image and sense of self. Evolution of body image begins as a child begins to interact with objects in the environment.

As children grow older, parents touch them less then they did during infancy and childhood. By adolescence, touch by parents may be nearly terminated. Yet adolescents become avid for tactile contact, seeking to touch and be touched. They are sexually mature, and this creates the dilemma.

Each year in the United States, more than 1 million girls under 20 become pregnant, and approximately 560,000 bear live children. Prevention of pregnancy is the ideal; contraception is one strategy, another is to encourage abstinence. However, some studies have indicated that because being held or cuddled reduces anxiety, promotes relaxation and a feeling of security, sexual intercourse may be used by adolescent girls as a way to be held and cuddled.

Therapeutic Touch

Being cuddled can help overcome feelings of loneliness and can generate feelings of love, reassurance, protection, and comfort. In a study on depression and the wish to be held, researchers found differences between females and males, however: Females at all levels of depression wished to be held, whereas the males who were the most depressed reported the greatest desire to be held.

Therapeutic touch, a modern derivative of the laying on of hands that involves touching with the intent to help or heal, was introduced into nursing in 1975 by Dolores Krieger. Krieger had observed that the treatment often seemed to help an ill person feel more relaxed, comfortable, and energetic. She wrote that "an exchange of vitality occurs when a healthy person purposefully touches an ill person with a strong intent to help or to heal." Therapeutic touch is learned: She describes the individual administering therapeutic touch as assuming a meditative state and placing her hands on or close to the body of the person she intends to help. She then gently attunes to the patient, placing hands over areas of tension and redirecting those energies. How to study such a concept? Kreiger chose hemoglobin levels as an index of comparison and found some increase in the levels of patients treated with therapeutic touch compared with those of a control group. Therapeutic touch is still being investigated. One study with patients who had tension headache pain showed that 90 percent of the subjects treated with therapeutic touch experienced pain reduction. An average 70 percent pain reduction was sustained over four hours following therapeutic touch, twice the average reduction of pain following untrained placebo touch.

Touching has been found to have a physiological effect even when the patient is unconscious. One researcher found significant heart rate change during pulse taking or hand holding of curarized patients in a shock-trauma unit. Another found changes in blood pressure, heart rate, and respiratory rate in acutely ill patients whose hands had been held by nurses for up to three minutes.

Gratuitous Touch

A recent experiment put the touch on more than 100 diners at two restaurants. Researchers found that diners who had been touched on the hand or shoulder by the waitress tipped better than those who had not been touched at all. In spite of the bigger tips, the diners did not give more positive reviews of the restaurants or the waitress. A variation on this experiment showed higher tips were given when the waitress touched the female in a male-female couple.

Other experiments have shown that in vocational counseling sessions, counselors were rated more highly and perceived as more expert when they touched their clients. Touching has also been found to increase compliance to requests made by the toucher. Students who were "accidentally" touched by library clerks when their library cards were being returned gave higher ratings to both the clerks and the library than did a control group of students who were not touched.

Touch may be an effective teaching tool: A study of 171 male and female college students showed that those who were touched by their instructor during individual conferences gave their instructors higher ratings; in addition, those who were touched showed superior performance on the next course examination, scoring .58 standard deviations higher than the untouched students.

When a person is feeling good, perhaps it is easier to reach out. Studying touching behavior in competitive sports, researchers have found that winners gave and received significantly more touches than did losers. Most touches were on the hands, back, or shoulders.

We have certain tactile no-no's, of course, such as those against touching strangers or inflicting pain. But researchers have identified some more subtle taboos. Subjects reacted strongly to a touch coupled with a negative statement, such as a slap on the rear while commenting about someone's extra poundage. Kissing someone's hands while they're typing is not appreciated, nor is a massage to sunburned shoulders. People don't like to be startled by sudden touch or to be moved out of the way like chess pieces.

Some research has indicated that women respond more positively to touch than do men, but the studies of gender differences in touching are complicated by lack of knowledge about power-connoting versus friendly touches.

To Be Is to Touch

When I was born, I had two teenage sisters, so I have often felt as though I had three mothers. As a child, I stood on a stool each morning while my mother braided my hair. When my grandmother was working alone at her quilts, I stood behind her and took her hairpins out and combed her long gray hair. When I was older, I held still while my sister plucked my eyebrows and painted my fingernails. We were grooming. My grandmother is gone now, and my mother. I still have my father and two sisters and various extended family to love, but we live far apart and so our contact is mostly verbal.

My life alone seems to demand at least the presence of a cat. Cruiser sleeps uninterested beside my keyboard as I write. At the grocery store, I commiserate with the always-agreeable people who are buying pet food. After all, we have our pets in common. We compare notes. When I must pace and think, I pick up the cat and scratch his ears. Cats, animal behaviorists say, have been so successful in exploiting humans because they release the cuddling and stroking behavior humans display toward infants; dogs, on the other hand, release playmate behavior. Who would I talk aloud to, and who would I talk silly to, I sometimes think, if I didn't have the cat? And who would jump on the bed and curl up at the small of my back on winter nights if not the cat?

A psychologist tells me that older people who have lost their mates live longer if they have their family nearby. I can recommend pets, too, for giving one a reason to live. Even the less cuddly ones have things to offer—another life. The Chinese kept crickets on the hearth and buried them in little silver caskets; a hydrophilid beetle in a fishbowl brought me great pleasure for almost a year (I fed it bits of liver). Modern society has dispersed, fragmented, and shrunk our extended family. I find that living alone, I seek out more sociality, as though I need a base of ten to replace a husband or a grown child or a parent or a sister. When I am with friends, I notice that we unselfconsciously hug when we meet and hug when we say goodbye.

What saves me from the slough of despond and keeps me going is a delight in the senses, all five of them, and first and most compelling of all, an undying (so far) curiosity.

—Louis Untermeyer

Variety is not the spice of life. It is the very stuff of it.

—Christopher Burney

We know nothing except by experience, and experience consists of nothing but the information of our senses. Perhaps there is nothing, really, out there to be sensed. All we know is that we sense.

—Judson Jerome

Remember all paydays of lilacs and songbirds.

—Carl Sandburg

The metaphors of mind are the world it perceives.

—Julian Jaynes

Good morning, good morning, the sun is up high
And no one on Earth is as happy as I.
The robins are singing, up high in the trees
And out in the garden are busy brown bees.

—My mother

S · E · V · E · N

Seeking Sensory Experience

Human Versatility

When I am standing on the balcony in the fieldhouse waiting for my racquetball partners, I see all sizes and shapes and colors of humans running around the track. Males and females. Some are talking. Many are wearing portable tape players with earphones.

In the center of the track I see an aerobics class, outfitted in tight, brilliant scraps of cloth, stretching, leaning, twisting, dancing together to music. I also see tennis players, outstretched like Leonardo figures, like the homunculus, their hands made even larger and longer by the racket. Whack, whack. They lean over and touch their toes. Then they stretch again, rotate an arm to whack the ball with precision. At another end of the fieldhouse, I see basketball players, wet muscles rippling, hunching, dribbling, dodging each other, then reaching roundly, fingers spread, to put the ball through the hoop. They smile, pleased with themselves. In an alcove on the side are the weightlifters, grimacing, grunting, glistening. Is our history in doubt?

The place smells, but it is not an unpleasant smell.

It is the smell of naked apes in Spandex.

I see and hear and smell these people, and I can squint my eyes, and I can imagine a yellow savanna, and a troop of curious social humanoids waving arms, jumping around, preening, communicating, sniffing each other, bonding, one-upping, establishing social dominance.

Now they glisten with exertion. Later, I imagine they will yawn and rest a little. Maybe they will climb one of the scattered trees (or sit on a bench) and eat an ice cream cone, overlooking the artificial grasslands we work so hard to maintain.

In the fieldhouse I see some of the ways we exercise our senses and synthesize our sensory information and motor skills—input, output. (The best among us are groomed for professional performance, while others watch.)

But I recognize these people at the university, and they are people, I know, who also teach quantum mechanics and chemistry and languages, who sit at computers and tell them what to do, who peer through microscopes, who study viruses and manipulate the genes of plants, who design dams for taming water, and who design and build satellites to be shot into space.

Our human engineering design—our flexible bodies combined with our sensitivities—makes us extremely versatile.

General Sensitivity

My friend Susan is bookish and subdued. She speaks softly and enjoys quiet activities, such as writing, drawing, and gardening. She notices details of everyday things, such as the songs of birds and the opening of flowers and the sounds leaves make in the wind. She is compassionate and a good listener. She needs time and space to think. She seems to know when people are sad, and she writes lovely notes to them.

My friend Leona has a high energy level. She drops by suddenly, pounds on the door, and says, "Let's go do something. Let's have fun." Her friends marvel at her, and we let her organize us. She seeks a lot of physical activity and stimulation. She rides 30 miles on a bicycle as though it were nothing. She does not hesitate to take apart a broken appliance to fix it. She's in town. She's out of town. Lost loves don't bother her for long. Life is too short. "Intellectualize it, and the emotions follow," she says. She's always ready to try something new. Every day seems filled with new stimulation.

Most of us fall somewhere in between in seeking stimulation, but as part of our common language, we talk about "sensitive" people or shy people and about people who are "outgoing" or risk takers, who want to push themselves or frighten themselves to feel stimulated.

Although our environment, experience, age, memory, and conditioning have their roles to play, our neurobiology must take a great deal of responsibility for our stimulus-seeking and activity levels, and our tastes in food, odors, music, art, and in other people.

Studies have shown that taste sensitivity and drug sensitivity are manifestations of a general kind of systemic sensitivity, and these have been correlated with the personality characteristics of introversion and extroversion. For instance, people identified as extreme introverts secrete much more saliva when stimulated by lemon juice than do extreme extroverts. Extroverts seem to have higher thresholds for other sensory experiences, requiring a greater intensity of stimulus to perceive stimulation.

When the extroverts are given an electroencephalogram (EEG), they show brain wave forms that indicate low levels of arousal. They are gregarious and are thought to need stimulation to stay alert. Arousal can also be measured as the ability to pay attention without distraction, and extroverts are usually not as good at tasks requiring stable attention. EEGs have also shown that extroverts are more affected by hallucinogenic drugs, that they dream infrequently and in black-and-white rather than in color. Introverts, on the other hand, dream more in color, prefer color to form, and display high energy content and greater variability on EEG measures of brain waves.

Other differences related to the senses have been found as well: Extroverted people smoke more cigarettes, drink more coffee and alcohol, consume more sugar, eat more and spicier foods, have sexual intercourse more frequently and from an earlier age, and generally seek stronger sensory stimulation than do introverts. Extroverts show greater tolerance for pain, less tolerance for stimulus deprivation, and shorter aftereffects of perceptual experiences. Introverts show greater arousal of the cerebral cortex, whereas extroverts habituate more quickly to stimuli.

Do opposites attract? Well, of course, males and females do. But I once met a man who spoke slowly, and he told me that he liked my "slow, measured voice." As I became acquainted with him, I learned that we shared many physical and personality traits. Narcissistically, I liked him. I felt I was confronting another manifestation of some of my ancestors' European genes, many of which had come together by chance in another person, a nonrelative. I was sensitive to him.

At the beginning of a new male-female relationship, human senses seem to be turned up. We're giddy. In *The Chemistry of Love*, Michael Leibowitz likens this phase to an amphetamine phase. Personal space breaks down, so we can obtain information from a closer perspective. The increased sensitivity, perhaps, is a result of adrenalin produced with the emotion of falling in love. People who are getting to know each other, and like each other, are animated, and the animation contributes to the gathering of useful information about the other person. At the beginning of a new relationship, a male and female talk a great deal and are extremely interested in and tolerant of each other's egocentric conversation. During this time they also gather information about the other person's reaction to the information they provide about themselves. They compare schemas: likes, dislikes, attitudes, and tastes. Using all their senses, they examine each other physically, chemically, and emotionally, asking if their sensory worlds connect.

The result: Humans tend to choose partners they can count on, who resemble themselves or their close relatives and/or who have similar sensitivities.

Adaptation and Recovery

Then, after the bond is forged, the senses turn down. (Or, as Leibowitz discusses, they enter a kind of opiate phase, in which each other's presence has a calming effect.) Schemas have been established. Television and other life demands enter. Conversation time drops dramatically—some couples talk only four minutes a day.

A feature of all the senses is *adaptation*—the cessation of response to a stimulus after repeated exposure to it. We also talk about *habituation*—getting used to a stimulus so that we no longer react to it. Infants will look less and less at a stimulus that is presented again and again. Humans sleep soundly in a house next to a busy freeway. Pigeons feed alongside the same freeway, with cars whizzing by.

Habituation is adaptive. We don't waste time reacting to something that we've learned about, and so can reserve our reaction for something else, something new. Our senses are designed to be more alert to change than to constant stimulation. We smell a new perfume keenly at first. Then the nasal receptors make room for new smells. (My perfume has become a name for me and a smell for acquaintances. What are you wearing? they say.)

Taste buds adapt too. We overload on rum-raisin ice cream. Nothing is as good as the first bite. And we don't want it every day.

Sated, we must wait for sensory recovery. Recovery takes different times for different stimuli.

The visual senses habituate too. Traveling through field after field of sunflowers in Kansas or hour after hour of sagebrush in the Great Basin Desert is hypnotic, so different from traveling through the constantly changing canyonlands of Utah. We have made national parks, shrines, of special visual stimuli in our landscape. (But the people who live in the canyonlands, unfortunately for them, have visually habitutated.) Nothing is as good as the first gasping view of the Grand Canyon. Nothing is as good as the first view of an architectural marvel. Nothing is as good as that first view of an amoeba through a microscope, the moon through a telescope, or the Earth through binoculars.

Nothing is as fine as the first concert, the first large chorus of human voices. Nothing is as good as the first touch of that new kitten. Nothing is as fine as holding hands that first time, with that strange attractive person, nothing as exhilarating as the first kiss. We keenly remember our firsts, and we sing of them.

But—if we wait a while for our senses to recover, then we can experience that intense sensation again with pleasure.

Primates are curious creatures. We are long-lived and big-brained. We learn quickly about our environment, which helps us to manipulate it. Habituation sends us in pursuit of new and different stimulation. We seek change. We need a vacation from repetitive tasks to come back to them renewed. A honeymoon ends. A change is as good as a rest. A change, a respite, revs up the senses.

We seek out sensory nuance in the wine, in the food, in the musical harmonics. We detect something we didn't detect the first time. We see details of a painting. We create different romantic situations with the same person, for different stimuli. Erotica? Stimulation for flagging senses. We take up new hobbies. We move. We buy something new to decorate ourselves, our homes. We have a garage sale. We have a midlife crisis. We have a love affair.

Yet we also seek out stability, group-belonging, and we each find repeated pleasure in certain sensory stimulation. Those good memories again. A chocolate cake. A cup of tea. A bouquet of roses. A fistful of sharpened pencils. A good friend. We listen to music, go to the movies or shopping. We go out to eat or out for ice cream. We go to the circus or to the amusement park.

Feed Me, Feed Me: Superstimulation

Supernormal releasers of behavior have been well documented in the animal world. Just as an oystercatcher will choose to incubate a foolishly large fake egg over its own normal-size egg (and brood parasites such as cowbirds and cuckoos have taken advantage of this characteristic to get other birds to raise their offspring), we humans seek superstimuli for our senses.

The natural stimulus doesn't have to be the optimal stimulus.

We surround ourselves with ever-larger television sets, giant stereo speakers, amplifiers. We keep in touch with telephones, with answering machines. We produce animated cartoons, artificial landscapes, panavision, electronic music. We go to rock concerts, to nightclubs with flashing lights. Weber's law tells us that the bigger the stimulus, the bigger the change needed for it to be detectable ($10 enriches a pauper, but more money is required to make a difference to a millionaire).

We oversugar, oversalt. We stimulate our tongues with spices and artificial sweeteners, our noses with artificial flavors, our eyes with artificial brilliant dyes. These are not things that nature has made, necessarily, although she provided the raw materials. They are superstimulants that humans have made or found. We seek out brilliance. We take risks to stimulate our senses—skydiving, rock-climbing. Some of us seek sensory thrills more than others do, or in different ways. We are sensory addicts. We like life in the fast lane: swings, rollercoasters, cars, motorcycles, speedboats, airplanes, rockets.

We adorn ourselves with scents and jewelry and especially with color: clothing and a variety of paints. Fashion changes quickly. We find different ways to emphasize this body part, that body part—and the body parts that make us different, notably shoulders on males, breasts and buttocks on women. We are the same but different, new. See?

And yet, when we wish to rest, when we need to get away from it all, we seek out the greenery of the ancestral forests, the landscaping of the savanna, the soothing noise of rustling leaves or running water, quiet conversation, the rhythm of a nearby heartbeat. We may return to flotation tanks, like the waters of the womb or of our fishy past.

Of Spice and Space

A changing sensory environment is essential for us. Nothing is worse for a human than monotony, solitary confinement. Without sensory stimulation, scientists have found, hallucinations begin. The mind begins to create its own strange images, to stimulate itself in weird ways.

Those who overwinter in Antarctica are required to have a psychological test before they are approved to be socked in for the winter. Psychologists are studying personality types that might do well alone, in confined places, to reveal what kind of people might function best in the confines of prolonged spaceflight. The problem I see is this: Those people who would be inclined to be happy reading books are not necessarily the same people who might relish being shot into space. Space needs both Susans and Leonas.

Scientists are busily working on how to design a space station, a closed environment, a little terrarium, a little Earth. And they are trying to figure out who best to put in it. We, ourselves, the human factor, present the greatest challenge. We are Earth designs. Flexible, versatile, but physiologically and anatomically Earth designs nonetheless. Without gravity, we lose bone and muscle, we become jellyfish.

The scientists are also working on plants that might grow in zero gravity and in the light of space to provide the best foods for space travelers on long flights, and they hypothesize that the act of growing the plants might be soothing—stimulating—for the astronauts. But if the plants are grown in hydroponic solutions, the astronauts would have no earth smells to remind them of hopscotch. No tomato hornworms or corn earworms. No cabbage butterflies. But—people don't have these things in our cities, and they get along just fine, they tell me. Still I wonder: On the space station, what are the chances of glimpsing a chance spider building a web?

To satisfy the great human curiosity for learning more about our universe, to sense beyond what we sense now, scientists are trying to design a world in which people will be sensorily happy—and healthy, of course—while hurtling through space. Human curiosity is great, after all, and the natural stimulus doesn't have to be the optimal stimulus. The voyagers can take their tape players and television sets.

Space is the ultimate expedition to the poles, the ultimate voyage in search of spices, but—without the possibilities of polar bears or whales, without penguins or the seaweed of the Sargasso Sea.

This primate will stay behind on Earth, I say, with gravity acting on my semicircular canals as it did on those of my ancestors. I'll watch it all from this end on television. Then I'll go outside, where I can hear cricket sounds, where I might catch the buzz of a hummingbird or the chirp of a nighthawk. I'll stay here and grow a little thyme for flavor, a little lavender for smell, and some zinnias for color. I'll go for a walk and pick up a few smooth stones.

References

CHAPTER 1

Miller, J. A. 1985. Brain selects among sights and sounds. *Sci. News* 128:312.

Viereck, R. 1988. The biology of poetry. *Poets & Writers Magazine* 16:7-12, May/June.

Vonnegut, K. 1982. *Deadeye Dick*. New York: Dell Publ.

CHAPTER 2

Alfred, R. 1978. Saturday night earache. *New West* 3:NC-49(1), Dec. 4.

Askill, J. 1979. *Physics of Musical Sounds*. New York: Van Nostrand.

Aural root of paranoia. 1983. *Sci Digest* 91:90(a), Jan.

Barry, D. 1989. Another round of amazing medical news. *Deseret News* 51, Jan. 15.

Beadle, M. 1977. *The Cat*. New York: Simon & Schuster, Inc.

Bostrom, B., C. S. Bostrom, and J. A. Bostrom. 1983. Listen! (letter) *New England J. Med.* 309:1194, Nov. 10.

Brody, J. E. 1983. Personal health: New relief for victims of tinnitus, a persistent ringing in the ears. *New York Times* 129:C8, March 26.

Bruce, R. V. 1988. Alexander Graham Bell. *Natl. Geogr.* 174:358-384, Sept.

Calvin, W. H. 1983. *The Throwing Madonna: Essays on the Brain*. New York: McGraw-Hill Book Co.

Cartwright, F. F. 1972. *Disease and History*. New York: Thomas Y. Crowell Co.

DeLoughry, T. J. 1988. Protesters declare victory after trustees of university for deaf choose first hearing-impaired president. *Chronicle of Higher Ed.* A13, March 23.

Eddington, D. K. 1983. Speech recognition in deaf subjects with multichannel intracochlear electrodes. *Ann. NY Acad. Sci.* 241-258.

Engle, M. 1983. Electronic ear implant will allow deaf to hear. *Washington Post* 197:C11, Dec. 20.

Fantel, H. 1983. Warning lights flash for earphone users. *New York Times* 132, Section 2:H19(L), July 24.

Forsyth, P. 1984. Artificial implants and the gift of sound. *Stanford Medicine*, Fall.

Freese, A. S. 1982. Non-existent noise no joke to victims. *Los Angeles Times* 102:I-C2, Dec. 9.

Gannon, J. R. 1981. *Deaf Heritage: A Narrative History of Deaf America*. Silver Spring, MD: National Association of the Deaf.

Griffin, D. R. 1958. More about bat "radar." *Sci. Am.*, July.

Hearing: A link to IQ? 1976. *Newsweek* 87:97(a), June 14.
Huyghe, P. 1988. Voices from inner space. *Hippocrates*, Jul/Aug.
Kastor, E. 1983. Hail to the chief for lending an ear. *Washington Post* 106:E5, Sept. 15.
Kolata, G. 1988. Tendency to get ear infections is inherited, study shows. *New York Times*, Feb. 16.
Komroff, M. 1961. *Beethoven and the World of Music*. Westport, CT: Greenwood Press.
Lohr, S. 1982. Headsets and ear damage. *New York Times* 131:12(L), July 17.
Magid, J. 1988. Psychiatrist suspects inner ear could be the culprit. *Salt Lake Tribune* A:11, June 14.
McIntyre, J. 1974. *Mind in the Waters*. New York: Charles Scribner's Sons.
Middle ear and learning. 1979. *Ed. Digest* 54:66(2), Feb.
Miller, M. H., and C. A. Silverman (eds.). 1984. *Occupational Hearing Conservation*. Englewood Cliffs, NJ: Prentice-Hall, Inc.
Milne, L., and M. Milne. 1962. *The Senses of Animals and Men*. New York: Atheneum.
Moore, B. C. J. 1982. *An Introduction to the Psychology of Hearing*, 2d Ed. New York: Academic Press.
Noble, W. G. 1978. *Assessment of Impaired Hearing: A Critique and a New Method*. New York: Academic Press.
Noises in the ear. 1981. *New York Times* 130:C3(LC), March 24.
Novacek, M. J. 1988. Navigators of the night. *Nat. Hist.* 97:67-70, Oct.
Paranoia cause. 1981. *U.S. News & World Report* 91:64(1), July 27.
Rader, C. M., and Tellegen, A. 1987. An investigation of synesthesia. *J. Personal. Soc. Psychol.* 51:981-987.
Roeder, K. D. 1965. Moths and ultrasound. *Sci. Am.* 212:94-102, April.
Rovner, S. 1983. Treating preschool ear problems: Tube has become a mainstay in avoiding hearing loss. *Los Angeles Times* 103:V5, Dec. 27.
Sataloff, J. 1966. *Hearing Loss*. Philadelphia: J. B. Lippincott Co.
Schauffler, R. H. 1929. *Beethoven: The Man Who Freed Music*. New York: Doubleday, Doran & Co.
Severo, R. 1981. 'Paranoia' in elderly attributed to unrecognized hearing loss. *New York Times* 130:C1(LC), June 23.
Shilkret, R. 1988. Letter to the editor. *Poets & Writers Magazine*, Oct.
Skelton, G. 1983. 44 years after injury, President gets hearing aid. *Los Angeles Times* 102:I1,8, Sept. 8.
Sonnenschein, M. A. 1987. Financing a cochlear implant. *Shhh*, Mar/Apr.
Stebbins, W. C. 1980. The evolution of hearing in the mammals. In Popper, A. N. and R. R. Fay, *Comparative Studies of Hearing in Vertebrates*. New York: Springer-Verlag.
Success for the 'Bionic Ear.' 1984. *Time* 123:60, March 12.
Thurber, J. 1966. University days. In Burhans, C. S. Jr., *The Would-be Writer*. Waltham, MA: Blaisdell Publ. Co.
VonBekesy, G. 1957. The Ear. *Sci. Am.*, Aug.
We hear you, Mr. President (editorial). 1983. *New York Times* 132:Section 4, E24(L), Sept. 11.

Weisman, S. 1983. Reagan begins to wear a hearing aid in public. *New York Times* 132:A14(L), Sept. 8.
Wilker, D., 1989. The loud decibels of rock can destroy hearing. *Salt Lake Tribune* 8E, Jan. 15.
Williams, J. 1983. The President begins wearing a hearing aid. *Washington Post* 106:A3, Sept. 8.
Wilson, F. R. 1986. *Tone Deaf and All Thumbs? An Invitation to Music-making for Late Bloomers and Non-prodigies.* New York: Viking.

CHAPTER 3

Abell, G. O., D. Morrison, and S. C. Wolff. 1987. *Exploration of the Universe*, 5th Ed. New York: Saunders College Publishing.
Bauer, L. O., B. D. Strock, R. Goldstein, J. A. Stern, and L. C. Walrath. 1985. Auditory discrimination and the eyeblink. *Psychophysiology* 22:636-641.
Blest, A. D., and M. Carter. 1987. Morphogenesis of a tiered principal retina and the evolution of jumping spiders. *Nature* 328:152-155.
Collins, E. 1987. Eyeing myopia: Research suggests how reading could lead to nearsightedness. *Sci. Am.* 257:36, Oct.
Eibl-Eibesfeldt, I. 1975. *Ethology: The Biology of Behavior,* 2d Ed. New York: Holt, Rinehart & Winston.
Eye of the (emotional) storm. 1987. *Sci. News* 131:40, Jan.
Frey, W. H. 1985. *Crying: The Mystery of Tears.* Minneapolis, MN: Winston Press.
Gertsch, W. J. 1979. *American Spiders,* 2d Ed. New York: Van Nostrand Reinhold.
Goldstein, R., L. C. Walrath, J. A. Stern, and B. D. Strock. 1985. Blink activity in a discrimination task as a function of stimulus modality and schedule of presentation. *Psychophysiology* 22:629-641.
Gregory, R. L. 1973. *Eye and Brain: The Psychology of Seeing,* 2d Ed. New York: World University Library, McGraw-Hill Book Co.
Hess, E. H. 1965. Attitude and pupil size. *Sci. Am.* 212:46-54, April.
Hilbert, D. R. 1987. *Color and Color Perception: A Study in Anthropocentric Realism.* Stanford, CA: Center for the Study of Language and Information.
Hubel, D. H. 1988. *Eye, Brain and Vision.* New York: Scientific American Library.
Kirn, T. F. 1987. Ophthalmologists discuss methods to help physicians see what patients can't see. *JAMA* 257:1027-1028.
Kolata, G. 1988. Blindness of prematurity unexplained. *Science* 231:20-22, Jan 3.
Kolata, G. 1988. Glaucoma and cataract: Closing in on causes. *New York Times* Cl, July 12.
Leary, W. 1988. Test shows freezing is best treatment for infant eye disease. *New York Times,* Mar. 31.
Lindberg, D. C. 1976. *Theories of Vision from Al-Kindi to Kepler.* Chicago: Univ. Chicago Press.
McAuliffe, K. 1985. Visions: how do cats, fish and snakes see the world? *Omni* 7:50-59, July.
McDermott, J. 1985. Researchers find there is more to vision than meets the eye. *Smithsonian* 16:96-107, April.

Mueller, C. G., M. Rudolph, and the Editors of Life. 1966. *Light and Vision*. New York: Time Inc.

Pace, J. 1987. What behavioral optometrists see. *Consumers' Res. Mag.* 70:11-15, April.

Patterson, E. C. 1970. *John Dalton and the Atomic Theory*. New York: Anchor Books, Doubleday & Co.

Salvini-Plawen, L. V., and E. Mayr. 1977. On the evolution of photoreceptors and eyes. In M. K. Hecht, W. C. Steere, and B. Wallace (eds.), *Evol. Biol.*, Vol. 10. New York: Plenum.

Stern, J. A., L. C. Walrath, and R. Goldstein. 1984. The endogenous eyeblink. *Psychophysiology* 21:22-33.

Tolbert, M., and R. E. Lippman. 1987. Are your contact lenses as safe as you think? *FDA Consumer* 21:16-19, April.

Wald, G. 1950. Eye and Camera. *Sci. Am.*, Aug.

Wallman, J, M. D. Gottlieb, V. Rajaram, and L. A. Fugate-Wenzek. 1987. Local retinal regions control local eye growth and myopia. *Science* 237:73-77.

Weck, E. 1987. Taking a look at eye exams. *FDA Consumer* 21:14-17, May.

CHAPTERS 4 AND 5

A new nose. 1988. *Sci. Impact* 1, May.

Bedichek, R. 1960. *The Sense of Smell*. New York: Doubleday & Co.

Berglund, B., and T. Lindvall. 1982. Olfaction. Ch. 11, in *The Nose: Upper Airway Physiology and the Atmospheric Environment*, D. F. Proctor and I. Andersen, eds. New York: Elsevier Biomedical Press.

Best, D. 1986. Redesigning food products for an aging population. *Prepared Foods* 155:181-189, Oct.

Birch, L. L. 1980. Effects of peer models' food choices and eating behaviors on preschoolers' food preferences. *Child Dev.* 51:489-496.

Birch, L. L, and D. W. Marlin. 1982. I don't like it; I never tried it: Effects of exposure on two-year-old children's food preferences. *Appetite: J. Intake Res.* 3:353–360.

Blakeslee, S. 1988. Pinpointing the pathway of smell. *New York Times* C1, Oct. 4.

Block, I. 1933. *Odoratus sexualis*. New York: American Anthropological Society, Inc.

Breast cancer and sense of smell. 1985. *Sci. News* 128:153.

Budiansky, S. 1986. Siren song of the pheromones. *U.S. News & World Report* 64, Dec. 1.

Byrne, G. 1987. Get a whiff of these data. *The Scientist*, Nov. 2.

Cain, W. S. 1981. Educating your nose. *Psychol. Today* 48-50, Jul.

Camazine, S. 1985. Olfactory aposematism: Association of food toxicity with naturally occurring odor. *J. Chem. Ecol.* 11(9):1289-1295.

Carterette, E. C., and M. P. Friedman. 1978. *Handbook of Perception: Vol. VIA Tasting and Smelling*. New York: Academic Press.

Comfort, A. 1971. Communication may be odorous. *New Scientist & Sci. J.* 49:412-414, Feb. 25.

Coppersmith, R., and M. Leon. 1984. Enhanced neural response to familiar olfactory cues. *Science* 225:849, Aug. 24.

Corbin, A. 1986. The foul and the fragrant: Odor and the French social imagination. *NY Rev. Books* 33:24-26, Nov. 20.

Cutler, W. B., G. Preti, A. Krieger, G. R. Huggins, C. R. Garcia, and H. J. Lawley. 1986. Human axillary secretions influence women's menstrual cycles: The influence of donor extract from men. *Hormones and Behavior* 20:463-473.

Davis, J. 1984. Smell lab: How do the nose and tongue transmit the data of wine and roses? *Omni* 6:44-46,96-98, Jan.

Dethier, V. G. 1962. *To Know a Fly*. San Francisco: Holden-Day.

Diggs, M. 1988. Hot, spicy food creations growing in popularity. *Herald Journal*, Logan, UT, 18, Oct. 19.

Dodge, B. S. 1988. *Quests for Spices and New Worlds*. Hamden, CT: Archon Books.

Doty, R. L. 1986. Cross-cultural studies of taste and smell perception. In *Chemical Signals in Vertebrates*, D. Duvall, ed. New York: Plenum.

Doty, R. L. 1986. Reproductive endocrine influences upon olfactory perception: A current perspective. *J. Chem. Ecol.* 12:497-509.

Doty, R. L., P. F. Reyes, and T. Gregor. 1987. Presence of both odor identification and detection deficits in Alzheimer's disease. *Brain Res. Bull.* 18:597-600.

Doty, R. L., P. Shaman, S. L. Applebaum, R. Giberson, L. Siksorski, and L. Rosenberg. 1984. Smell identification ability: Changes with age. *Science* 226:1441-1443, Dec. 21.

Doty, R. L., P. Shaman, and M. Dann. 1984. Development of the University of Pennsylvania smell identification test: A standardized microencapsulated test of olfactory function. *Physiol. & Behav.* 32:489-502.

Edmondson, B. 1986. National Geographic sells demographics. *Am. Demographics* 8:20, Dec.

Emsley, J. 1988. Artificial sweeteners. *ChemMatters* 6:4-8, Feb.

Engen, T. 1982. *The Perception of Odors*. New York: Academic Press.

Engler, N. 1985. Space taste. *Omni* 7:31, May.

FDA to study NutraSweet's new fat substitute. 1988. *Herald Journal*, Logan, UT, 4, Jan. 29.

Ferrara, J. L. 1987. Why vultures make good neighbors. *Natl. Wildl.* 16-20, Jun/Jul.

Fillion, T. J., and E. M. Blass. 1986. Infantile experience with suckling odors determines adult sexual behavior in male rats. *Science* 231:729, Feb. 14.

Forsyth, A. 1985. Good scents and bad: Odors can be a mammalian aphrodisiac or a turnoff. *Nat. Hist.* 94:24-31.

Genetics of odor-blindness. 1984. *Sci. News* 125:137, Sept. 1.

Gibbons, B. 1986. The intimate sense of smell. *Natl. Geogr.* 170:324-338.

Gilbert, A. N., and C. J. Wysocki. 1987. The smell survey: Its results. *Natl. Geogr.* 172:514-525, Oct.

Gilbert, S. 1984. Figuring out food preferences. *Sci. Digest* 12, Oct.

Goodhart, R. S., and M. E. Shils. 1980. *Modern Nutrition in Health and Disease*, 6th Ed. Philadelphia: Lea & Febinger.

Haagen-Smit, A. J. 1952. Smell and Taste. *Sci. Am.*, March.

Hasler, A. D., and A. T. Scholz. 1985. *Olfactory Imprinting and Homing in Salmon*. New York: Springer-Verlag.

Heil, J. 1985. The sexy scents of smell. *New Woman*, Feb.

Herbert, W. 1986. Sweet treatment. *Psychol. Today* 20:6-8, Dec.

Hodgson, E. W. 1961. Taste receptors. *Sci. Am.*, May.

Houston, D. C. 1986. Scavenging efficiency of turkey vultures in tropical forest. *Condor* 88:318-323.

Hundley, L. R. 1960. Taste test papers. *Carolina Tips*, June.

Kare, M. R., and J. G. Brand (eds.). 1986. *Interaction of the Chemical Senses With Nutrition*. New York: Academic Press.

Keverne, E. B. 1982. Chemical senses: Taste. Ch. 19, in *The Senses*, H. B. Barlow and J. D. Mollon, eds. New York: Cambridge Univ. Press.

Kim, S., A. deVos, and C. Ogata. 1988. Crystal structures of two intensely sweet proteins. *TIBS* 13:13-15, Jan.

Laddell, W. S. S. 1965. Water and salt (sodium chloride) intakes. Ch. 9, in O. G. Edholm and A. L. Bacharach (eds.), *The Physiology of Human Survival*. New York: Academic Press.

Lecos, C. 1985. Our insatiable sweet tooth. *FDA Consumer* 19:25, Oct.

Levin, D. A. 1976. The chemical defenses of plants to pathogens and herbivores. *Annu. Rev. Ecol. Syst.* 7:121-59.

Liebowitz, M. R. 1983. *The Chemistry of Love*. Boston: Little, Brown & Co.

Loveland, C. J., T. H. Furst, and G. C. Lauritzen. 1989. Geophagia in human populations. *Food and Food Ways*, in press.

McCarthy, P. 1986. Scent: The tie that binds? *Psychol. Today* 20:6-7, July.

McKinney, K. 1987. Nosy readers. *Omni* 9:18, May.

Mertens, T. R. 1988. Human genetics and McKusick's Mendelian Inheritance in Man. *Am. Biol. Teacher* 50:262-265, May.

Miller, J. A. 1984. Brain already busy while in the womb. *Sci. News*, Oct. 20.

Miller, J. A. 1984. Tour through a taste bud. *Sci. News* 126:383, Dec. 22,29.

Mishara, E. 1985. Robot nose. *Omni* 7:38, March.

Mobbe, G. 1985. Old noses. *Omni* 7:31, Aug.

Monell Chemical Senses Center. *Report*, 60 pp., summarizing research areas, and listing staff publications since 1986. Philadelphia.

Mowat, F. 1963. *Never Cry Wolf*. New York: Little, Brown & Co.

Newsome, R. L. (ed.) 1986. *Sweeteners: Nutritive and Non-nutritive. Scientific Status Summary by the Institute of Food Technologists Expert Panel on Food Safety & Nutrition*. Chicago: Institute of Food Technologists.

Ohloff, G. and A. F. Thomas (eds.). 1971. *Gustation and Olfaction*. New York: Academic Press.

Panati, C. 1987. *Extraordinary Origins of Everyday Things*. New York: Harper & Row.

Pevsner, J., R. R. Reed, P. G. Feinstein, and S. H. Snyder. 1988. Molecular cloning of odorant-binding protein: Member of a ligand carrier family. *Science* 241:336-339.

Preti, G., W. B. Cutler, C. M. Christensen, H. Lawley, G. R. Huggins, and C. Garcia. 1987. Human axillary extracts: Analysis of compounds from samples which influence menstrual timing. *J. Chem. Ecol.* 13:717-731.

Preti, G., W. B. Cutler, C. R. Garcia, G. R. Huggins, and H. J. Lawley. 1986. Human axillary secretions influence women's menstrual cycles: The role of donor extract of females. *Horm. Behav.* 20:474-482.

Probber, J. 1987. At Monell, it's all in the taste and smell. *New York Times* C1,C8, Nov. 25.

Rensberger, B. 1985. Noses know by means unknown: Scientists hunt olfactory facts. *Washington Post* D1, April 7.

Reyneri, A. 1984. The nose knows, but science doesn't. *Sci. '84* 5:26, April.

Robards, T. 1977. *The New York Times Book of Wine.* New York: Avon Books.

Roper, S. D., and J. Atema. 1987. Olfaction and Taste IX. *Ann. NY Acad. Sci.*, Vol. 510.

Roraback, D. 1987. Entrepreneurs market the ultimate: Human sex spray. *Salt Lake Tribune* A11, Nov. 1.

Roueche, B. 1983. A matter of taste. *Saturday Evening Post* 255:66-69, April.

Scherr, G. H. (ed.) 1980. *Saccharin: A Report by Dr. Morris F. Cranmer.* Pathotox Publ., Inc.

Selzer, Richard. 1979. Travels in rhineland. In *Confessions of a Knife.* New York: William Morrow.

Shell, E. R. 1986. Chemists whip up a tasty mess of artificial flavors. *Smithsonian* 17:78-88, May.

Sheraton, M. 1984. The critical palate, and You can argue with taste. *Time* 123-74-75, June 4.

Skryja, D. D. 1978. Reproductive inhibition in female cactus mice *(Peromyscus eremicus). J. Mammal.* 59:543-550.

Smith, L. 1982. Adventures in the sex and hunger trade. *Fortune*, Aug. 9.

Smith, S. A., and R. A. Paselk. 1986. Olfactory sensitivity of the turkey vulture *(Cathartes aura)* to three carrion-associated odorants. *Auk* 103:586-592.

Stegink, L. D., and L. J. Filer Jr. 1984. *Aspartame: Physiology and Biochemistry.* New York: Marcel Dekker, Inc.

Stone, J. 1988. Life-styles of the rich and creamy. *Discover* 9:81-82.

Süskind, P. 1986. *Perfume: The Story of a Murderer.* New York: Alfred A. Knopf.

There's no smell like home. 1984. *Sci. '84* 5:10-12, Jul/Aug.

Wasser, S. K., and D. P. Barash. 1983. Reproductive suppression among female mammals: Implications for biomedicine and sexual selection theory. *Q. Rev. Biol.* 58:513-538.

Weintraub, P. 1986. Scentimental journeys. *Omni* 8:48-52,114-116, April.

Wickler, W. 1968. *Mimicry in Plants and Animals.* London: World University Library.

Zamula, E. 1985. The curious compulsion called pica. *FDA Consumer* 18:29-32, Dec/Jan.

CHAPTER 6

Anderson, C. H., and R. V. Heckel. 1985. Touching behaviors of winners and losers in swimming races. *Perceptual and Motor Skills* 60:289-290.

Boffey, P. M. 1987. Infants' sense of pain is recognized, finally. *New York Times*, Nov. 24.

Brown, C. C. (ed.) 1984. *The Many Facets of Touch.* Pediatric Round Table: 10, Johnson & Johnson Baby Products Co.
Crusco, A. H., and C. G. Wetzel. 1984. The Midas touch: The effects of interpersonal touch on restaurant tipping. *Pers. Soc. Psychol. Bull.* 10:512-517.
Early stroking helps old age. 1988. *Sci. Impact* 1:8, Mar.
Finlayson, A. 1985. The healing touch. *Maclean's* 98:68, Dec. 9.
Geldard, F. 1984. The mutability of time and space on the skin. *Physics Today* S38-39, Jan.
Gibson, J. 1984. Hugs and kisses. *Parents* 140, April.
Goleman, D. 1988. The experience of touch: Research points to a critical role. *New York Times* C1, Feb. 2.
Gordon, G. (ed.). 1978. *Active Touch: The Mechanism of Recognition of Objects by Manipulation, A Multidisciplinary Approach.* New York: Pergamon Press.
Heckel, R. V., S. S. Allen, and D. C. Blackmon. 1986. Tactile communication of winners in flag football. *Perceptual and Motor Skills* 63:553-554.
Keele, K. D. 1957. *Anatomies of Pain.* Springfield, IL: Charles C. Thomas.
Keller, E., and V. M. Bzdek. 1986. Effects of therapeutic touch on tension headache pain. *Nursing Res.* 35:101-105.
Kruger, L., and J. C. Liebeskind. 1984. *Advances in Pain Research and Therapy, Vol. 6, Neural Mechanisms of Pain.* New York: Raven Press.
Marples, M. J. 1969. Life on the human skin. *Sci. Am.* 220:108-115.
McKean, K. 1986. Pain. *Discover* 82-92, Oct.
Montagu, A. 1986. *Touching: The Human Significance of the Skin,* 3d Ed. New York: Harper & Row.
Sachs, F. 1988. The intimate sense: Understanding the mechanics of touch. *The Sciences,* Jan/Feb.
Schiff, W., and E. Foulke. 1982. *Tactual Perception: A Sourcebook.* New York: Cambridge Univ. Press.
Schuman, W. 1984. Hugs & kisses. *Parents* 74-77, Nov.
Stein, N., and M. Sanfilipo. 1985. Depression and the wish to be held. *J. Clin. Psychol.* 41:3-9.
Stephen, R., and R. L. Zweigenhaft. 1985. The effect on tipping of a waitress touching male and female customers. *J. Soc. Psychol.* 126:141-142.
Steward, A. L., and M. Lupfer. 1987. Touching as teaching: The effect of touch on students' perceptions and performance. *J. Appl. Soc. Psychol.* 17:800-809.
Stier, D. S., and J. A. Hall. 1984. Gender differences in touch: An empirical and theoretical review. *J. Pers. Soc. Psychol.* 57:440-459.
Travis, M. 1985. A touch too much. *Psychol. Today,* 10, Feb.
Zotterman, Y. (ed.). 1976. *Sensory Functions of the Skin in Primates, With Special Reference to Man.* New York: Pergamon Press.

CHAPTER 7

Goleman, D. 1988. Why do people crave the experience? *New York Times* C1, Aug. 2.
Heron, W. 1957. The pathology of boredom. *Sci. Am.,* Jan.
Study exploring ways Native Americans coped with isolation. 1986. *Herald Journal,* Logan, UT, Feb. 5.

GENERAL

Alcock, J. 1975. *Animal Behavior: An Evolutionary Approach.* Sunderland, MA: Sinauer.

Boff, K. R., L. Kaufman, and J. P. Thomas. 1986. *Handbook of Perception and Human Performance, Vol. 1, Sensory Processes and Perception.* New York: John Wiley and Sons.

Bower, T. G. R. 1977. *The Perceptual World of the Child.* Cambridge, MA: Harvard Univ. Press.

Dawson, W. W., and J. M. Enoch (eds.). 1984. *Foundations of Sensory Science.* New York: Springer-Verlag.

Gibson, J. J. 1966. *The Senses Considered as Perceptual Systems.* Boston: Houghton Mifflin Co.

Goldstein, E. B. 1980. *Sensation and Perception.* Belmont, CA: Wadsworth.

Gross, C. G., and H. P. Zeigler. 1969. *Readings in Physiological Psychology: Neurophysiology/Sensory Processes.* New York: Harper & Row.

Levine, M. W., and J. M. Shefner. 1981. *Fundamentals of Sensation and Perceptions.* Reading, MA: Addison-Wesley.

Schmidt, R. F. 1978. *Fundamentals of Sensory Physiology.* New York: Springer-Verlag.

Sekuler, R., and R. Blake. 1987. Sensory overload: Understanding how senses age can help older people and their families fight back. *Psychol. Today* 21:48-51.

Siegel, R. E. 1970. *Galen on Sense Perception.* New York: S. Karger.

Terres, J. K. 1980. *Audubon Society Encyclopedia of North American Birds.* New York: Alfred A. Knopf, Inc.

Tortora, G. J., and N. P. Anagnostakos. 1984. *Principles of Anatomy and Physiology,* 4th Ed. New York: Harper and Row.

Uttal, W. R. (ed.). 1972. *Sensory Coding: Selected Readings.* Boston: Little, Brown & Co.

Wessells, N. K., and J. L. Hopson. 1988. *Biology.* New York: Random House, Inc.

The *Encyclopaedia Britannica*, 11th Ed., Cambridge, England: University Press, was useful for information that existed before 1911 and provided the definitions at the beginnings of the chapters.

Index

Abbe de l'Epee, 48
Adaptation, sensory, 145, 190, 192, 209
Ageusia, 171
Aglycogeusia, 171
Alcmaeon, 142
Alexander the Great, 127, 132
Allomones, 113
Amphioxus, 98
Anosmia, 124–125
Aqueous humor, 58, 83
Aristotle, 11, 60, 73, 106, 143, 144
Armato, Salvina, 80
Artificial sweeteners, 140, 146, 152–157, 175
Aspartame, 155–156
Astigmatism, 78–79
Attachment bonds, 196–197, 198
Audiograms, 36, 38, 39
Auditory agnosia, 44
Auditory nerve, 17, 21
Avicenna, 143

Bacon, Francis, 128
Baranski, S., 173
Bathing, 129–130
Becket, Thomas, 130
Beethoven, Ludwig van, 2, 44–45
Bell, Alexander Graham, 49
Better Hearing Institute, 41
Bitter taste, 136, 137, 139–140, 143, 145, 162–164, 165–166
Blind spot (optic disk), 58, 59
Blink science, 90
Bombykol, 114
Bombyx mori, 114–115
Bothrops jararaca, 194
Bradykinin, 193–194
Braille, Louis, 188
  and tactile alphabet, 188–189
Bruce effect, 112
Bruce, Hilda, 113

Calvin, William, 33
Cartwright, F. F., 43
Cataracts, 84
Cat in the Hat, The, 33
Chemical communication, 7, 94–95, 100, 173–174
Chemistry of Love, The, 208
Children of a Lesser God, 47
Chorioretinopathy, acute, 83
Choroid, 55, 56
Ciliary muscle, 55, 56, 58
Ciliary nerve, 55
Cinchona, 163
Clark, Tom, 34
Cleopatra, 131, 132
Clerc, Laurent, 48
Cochlea, 16, 17, 20–22, 23, 38, 41, 50
Cochlear implants, 50–51
Cocktail party effect, 38
Code on the Eye, 81
Cold receptors, 189–191
Color blindness, 70–73
Comfort zone, temperature, 190
Compact discs (CDs), 40
Conchae, 96
Cones, 57, 64, 69, 70
Coniglio, Adolph, 171–172
Contact lenses, 81–82

Cornea, 55, 73, 78, 85
Corneal transplant, 85
Corn syrup, 150–151
Cribriform plate, 95–96
Crying: The Mystery of Tears, 87
Cuckoo–pint (*Arum maculatum*), 104–105
Cyclamate, 154, 157

da Gama, Vasco, 176–177
Dalton, John, 71–73
da Vinci, Leonardo, 81
Deafness. *See also* hearing loss
- hereditary, 43–44
- nerve, 42

Decibel levels, 37
- and hearing loss, 41

de'Medici, Giovanni Cardinal, 80
Democritus, 128, 142, 144
Der Aufbau der Tastwelt, 186
Dermis, 181
Descartes, René, 73
Dethier, Vincent, 174–175
Deuteranopes, 71
Diabetic retinopathy, 83
Dichromats, 71
Dr. Seuss, 33
Dysgeusia, 171
- idiopathic, 171, 172

Ear, 16–24
Eardrum. *See also* tympanic membrane
- aging of, 42

Ear tubes, 35
Earwax, 18, 41
Eibl–Eibesfeldt, Irenaeus, 89
1812 Overture, 30
Electrocutaneous devices, 49
Electromagnetic spectrum, 59, 61
End bulbs of Krause, 182, 183, 184
End organs of Ruffini, 182, 183, 184
Epidermis, 181
Eskimo, 79, 189
Ethmoid, 95
Euglena, 65
Eustachian tube, 17, 20, 34–35
Evolution, 6, 22–24, 66, 70, 98, 105
Eye, 54–59, 65–70
- and camera compared, 73–75
- pain receptors, 192

Eyebrow flash, 89
Eyeglasses, 80–81

Fats, 164–165
Fechner, Gustav, 143
Fernel, Jean, 143
Fight or flight, 39, 192, 194
Floaters, 59
Fovea, 56, 57, 62, 69
Franklin, Benjamin, 81
Franklin, Jon, 75
Frey, M. von, 183
Frey, William H. II, 87
Fruit flies (*Drosophila*), 29–30, 121

Galen, 143, 186
Galileo, 11, 86
Galopin, Augustin, 128
Galvani, Luigi, 143
Gaullaudet School, 48–49
Gaullaudet, Thomas, 48
Glands of Zeis, 55
Glaucoma, 83, 84
Griffin, Donald, 28
Grooming, 129, 196, 197
Guinea Pig Doctors, 75
Gymnema sylvestre, 141

Habituation, 209, 210
Hahn, Helmut, 143–144
Hands, 186–188
Harlow, Harry F., 196
Hearing, 15–51
- in barn owl, 5, 30–31
- in bats, 23, 27–29
- in cats, 36
- in children, 19, 35
- in moths, 28–29
- in whales and dolphins, 23, 32
- mammalian, 23

Hearing loss
  and drugs, 42, 43
  and headphones, 41
  and paranoia, 47
  conductive, 41–42
  industrial, 38
  in high frequencies, 19
  temporary, 38, 42
Heat receptors, 189–191
Helmholtz, Hermann von, 64, 80, 143
Henkin, Robert I., 171–172
Henning, Hans, 107, 108
Hero of Alexandria, 60
Hess, Eckhard, 88–89
Hippocrates, 118
Holmes, Oliver Wendell, 181
Homer, 161
Homo sapiens, 10
Homunculus, 12–13, 206
Honey, 147–149
Hooke, Robert, 85
Hubbard, Mabel, 49
Hurt, William, 47
Huygens, Christian, 60
Hyperopia, 77–78
Hypogeusia, 171, 172
Hyposmia, 124

Incus, 17, 20
Inner ear, 20, 21. *See also* cochlea
International Flavors and Fragrances, 132
Iris, 55, 56, 73–74

Jacobson's organ. *See* vomeronasal organ
Jaynes, Julian, 43
Joan of Arc, 42–43

Katz, David, 186
Keller, Helen, 127, 186
Kepler, Johannes, 73
Kim, Sung-Hou, 156
Koss, Michael, 39
Krieger, Dolores, 200
Kuhne, Willy, 74

Labyrinthine disease, 42
LeCat, Claude-Nicholas, 141
Leeuwenhoek, Antoni van, 27, 85
Leibowitz, Michael, 208, 209
Lens, 56, 58, 73, 74, 78, 84
Le Parfum de la Femme, 128
Linnaeus, Carolus (Karl Linne), 107
Louis XI, 124
Luther, Martin, 43

Macula lutea, 57, 74
Macular degeneration, 84
Magellan, Ferdinand, 158
Malleus, 17, 20
Matlin, Marlee, 47
Meibomian glands, 55, 56
Meissner's corpuscles, 182, 183, 184
Mendelian Inheritance in Man, 124
Meniere's disease, 42
Menstrual synchrony, 116–117
Mercaptan, 103–104, 109, 119, 124
Merkel's discs, 182, 184
Microscopes, 85–86
Middle ear, 20, 34–35
Middle ear reflex, 38
Minnich, D. E., 174
Monell Chemical Senses Center, 119, 176
Monellin, 156
Montagu, Ashley, 198–199
Morris, Desmond, 197
Morse, Samuel F. B., 49
Musk, 111, 112, 131
Myopia, 77–80

Nasal septum, 96
National Geographic smell survey, 119–120
Natural selection, 9
Neophobia, food, 142
Nero, 131

Newton, Sir Isaac, 60, 63, 64
Night blindness, 63
Nixon, Richard, 90
Noble, W. G., 48
Noise, 19–20
Noise notch, 38, 39
Nose, 94, 95, 96, 97
  artificial, 126

Odor blindness, 123–124
Odor deficits, 124
  in Alzheimer's disease, 124
  in Parkinson's disease, 124
Odors
  environmental, 126–127
  of illness, 128–129
Odyssey, 161
Olfaction
  and protein carriers, 111
  in bats, 102
  in carnivores, 100–101
  in gila monster, 101
  in rodents, 100–101
  in turkey vulture, 102–103
  stereochemical model of, 110
  vibrational model of, 110–111
Olfactory bulb (lobe), 95, 96, 97
Olfactory receptor cells, 95, 96, 97, 109
Ommatidia, 66
Omni smell survey, 121
Ophthalmoscope, 64, 80
Optical illusions, 75–76
Optic chiasma, 57
Optic nerve, 56, 57, 58
Optimality theory, 9
Optograms, 74
Origin of Consciousness in the Breakdown of the Bicameral Mind, The, 43
Ossicles, 16, 17, 20, 37,
Otitis media, 34–35
Otosclerosis, 42

Pacinian corpuscle, 182, 183
Paget's disease, 45
Pain, 181, 182, 183, 191–195
  acute, 195
  chronic, 194–195
  gate-control theory of, 193
  inhibition, 193
  in newborns, 192
  insensitivity, 195
  referred, 195
Pain receptors, 192
Pantogeusia, 171
Papillae, tongue, 137–138
  filiform, 138
  foliate, 137–138
  fungiform, 137
  vallate, 137
Parmenter, Bob, 102, 105
Pavarotti, Luciano, 24, 26
Perfumery, 130–132
Perfume: The Story of a Murderer, 94
Phenylketonuria, 128, 155
Pheromones, 112–115
  in ants, 113
  in honeybees, 113
  in pink bollworm, 114
  in rhesus monkey, 115
  in silkworm moth, 114
  in termites, 114
Physiological Optics, 64
Pica, 158
Pinkeye, 85
Pinnae, 17
Pitch, 25, 37
Pit eye, 65
Plato, 106, 161
Poppaea, 131
Preadaptations, 22
Prenatal school, 33
Presbycusis, 35, 36
Presbyopia, 58, 79
Pressure sensation, 181
Protanopes, 71
Psychophysics, 143–145
PTC, 165–166
Pupil, 55
  size and emotional response, 88–89
Pythagoras, 60

Quinine, 163, 166–167

Radial keratotomy, 82
Reagan, Ronald, 45
Retina, 54, 55, 56, 57, 62, 69, 74, 75, 78
  detached, 82
  dinosaur, 68
Retinopathy, 83, 84
Rhodopsin, 62–63, 74
Rods, 57, 62, 69

Saccharin, 153–155, 157
Salt
  need-free appetite for, 161
  physiological role of, 159
Salty taste, 136–137, 139, 144, 159–160
  and charged ions, 159
Sclera, 55, 56
Scratch-and-sniff, 118–119
Scurvy
  and vitamin C, 158, 177
Semicircular canals, 22
Senile miosis, 84
Sex smells, 117–118
Sexual attractants, animal, 131
Sign language, 47–49
Skin, 181–185
  as hearing aid, 49–50
Skin receptors, 184, 185
Skylab, 91, 172–173
Smell, 93–133. *See also* olfaction
Smell memories, 99, 120, 121–123, 125
  in rats, 99
  in ridley sea turtles, 99
  in salmon, 99
Smells, classification of, 106–109
Smoking
  and smell, 120
  and sugar consumption, 151
  and taste, 167, 168
Snell, Willebrord, 60
Sound waves, 18, 19
Sour taste, 136–137, 139, 157–158
Spallanzani, Lazzaro, 28
Speakes, Larry, 46
Spices, 176
Spinal meningitis, 34, 42, 83
Stapes, 17, 20, 21
Stern, John, 90
Stinkhorn (*Phallus impudicus*), 105
Sty, 85
Sugar, 146–147, 149–152
  alcohols, 152
  and mood. 151–152
Sugarbeet, 149–150
Sugarcane, 149–150
Sunglasses, 81
Supernormal releasers, 211
Superstimulants, 211
Süskind, Patrick, 94
Sweat glands, 181
Sweet taste, 136–137, 139, 140
  and food preferences, 142
  present at birth, 141
Synesthesia, 26–27

Talbot, Fritz, 199
Tapetum, 57
Taste, 135–177
  and food preferences, 166–167
  aversions, 175–176
  detection threshold, 140
  hedonic response, 158
  intensity, 140
  molecular connectivity theory of, 146
  water, 145–146
Taste buds, 137–139, 140
  adaptation of, 209
  and age, 141, 168
  extralingual, 141, 168
  in space, 172–173
Taste profile, 144, 145
Taste responses
  in blowfly, 174–175
  in vertebrates, 157
Taste threshold
  and drug sensitivity, 167, 168, 207
Tchaikovsky, 30
Tears, 87–88
Telescopes, 85, 86–87

Temperature sensation, 164, 181, 183, 189–191
Thaumatin, 156
Theophrastus, 106
Throwing Madonna, The, 33
Thurber, James, 54
Tickle belt, 50
Tinnitus, 38, 42, 43
Tongue-mapping, 136–137, 144
Tonometer, 83
Touch, 179–203
  active and passive, 180
  and adolescence, 200
  and development, 198–200
  and primates, 196
  and rats, 198
  and restaurant tipping, 201
  and vocational counseling, 201
  depression and the wish for, 200
  in competitive sports, 202
  taboos, 202
  therapeutic, 200–201
Trachoma, 85
Trichromatic theory, 63–64
Trichromatic vision, 70
Trigeminal nerve, 98, 193
Trimethylaminuria, 129
Tritanopes, 71
Tympanic membrane, 17, 18

U.S. Occupational Safety and Health Administration, 41

Vaginal secretions, 115–116
van Gogh, Vincent, 16
Vibrotactile devices, 49
Visible light, 59–62
Vision, 53–91
  in African mole rats, 68
  in anableps
  in bats, 68
  in bees, 66
  in birds, 69
  in leopard frog, 68
  in nautilus, 65–66
  in primates, 70
  in spiders, 67–68
  in trilobites, 66
  in whirligig beetles, 66
  photopic, 63
  scotopic, 62
Vital capacity, 24
Vitreous humor, 56, 59, 84
Vocal cords, 25
Vocal organs, 24
Voice, human, 24–26
Volta, Alessandro, 143
Vomeronasal organ, 101–102
Vonnegut, Kurt, 2

Ware, James, 79
Weber, E. H., 191
Weber's law, 211
White noise, 40
Whitman, Walt, 127
Wilson, Edward O., 94
Wine tasting, 168–170
Wood, John, 40
Wundt, Wilhelm, 143

Young, Thomas, 63, 64, 72

Zinc, 84, 158, 172
Zotterman, Yngve, 144
Zwaardemaker, Hendrik, 107